A bird does not change its feathers because the weather is bad.

A bird in the hand is worth two in the bush.

NATIONAL
GEOGRAPHIC
KiDS

Fly With Me

A **CELEBRATION** OF **BIRDS** THROUGH **PICTURES, POEMS, AND STORIES**

JANE YOLEN, HEIDI E. Y. STEMPLE,
ADAM STEMPLE, and JASON STEMPLE

NATIONAL GEOGRAPHIC
WASHINGTON, D.C.

Contents

OSPREY

BUDGERIGARS

Introduction

FOR THE LOVE OF BIRDS

There are millions of birds in almost every corner of our planet, in all shapes and colors. They live in woods and on mountains, on the desert sands and sandy beaches. They fly over meadows and farmlands, perch in cities and towns, live in frozen tundras, dense forests, and even in backyards. They sing in great mixed choruses at dawn, and smaller groups at dusk. Some of them are awake in the day, some in the night.

Birds are far older than humans, having been the inheritors of the great dinosaurs, many of which—we are only now discovering—were covered with feathers, not scales. Yet, most of us rarely see the birds around us. Or notice them. Or can distinguish one type of bird from another, either by sight or by sound. But they are there, in all their beauty.

As a family of birders, we love birds and strive to know as much about them as possible. And we have written this book to celebrate the beauty and majesty of birds with you. We hope the stories and songs and science and poems found on these pages excite and inspire you to know more about the beautiful birds in our world, too.

Jane Yolen, Heidi E.Y. Stemple,
Adam Stemple, and Jason Stemple

BLUE-AND-YELLOW MACAW

WHAT *is* A BIRD?

What makes a bird different from other living creatures?

WHAT DEFINES A BIRD?

There are many definitions.

"A creature with feathers and wings, usually able to fly."
—Cambridge Academic Dictionary

"A warm-blooded egg-laying vertebrate
distinguished by the possession of feathers, wings,
and a beak and (typically) by being able to fly."
—Oxford English Dictionary

"An animal covered in feathers, with two wings for flying and a hard pointed mouth called a beak or a bill. Birds build nests, in which female birds lay eggs."
—Macmillan Dictionary

Defining a bird is complicated. Lots of people have tried to do it.

But the characteristics being used to define birds don't always distinguish them from other species. For example, birds are warm-blooded. But so are mammals—and that includes humans. And we are not birds. Most birds can fly, but so can bats and bees. Birds have bills. But duck-billed platypuses have bills, too. Birds lay eggs, but so do snakes.

To come up with the perfect definition of a bird—and only a bird—it is important to think about all birds, from the tiniest hummingbird to the largest ostrich, from waddling penguins to soaring eagles. Then we should ask what they have in common with one another.

That means looking at each part of a bird, and noting which other animals share each particular feature. What is the one thing that only birds have? Feathers! No other living creatures on Earth have feathers.

Scientific Names

Binomial nomenclature is a system of categorizing all living things. In Latin, binomial refers to "two" and nomenclature to "naming." So, when we speak of birds by their Latin or Greek scientific names, we use two names: genus and species.

Genus is a group of related living things that includes one or more species. Animals or plants in the same genus share one or more common characteristics. Species is a group of related animals or plants that share related genes and can produce young animals or plants.

This system was first used back in the mid-1700s. Today, it is international, so every known animal, bird, plant, and other organism can be identified even if their common names are different in each language. The genus always begins with a capital letter and the species with lowercase. These names are also known as scientific names and are always written in italics. In this book you will see the scientific names of birds in facts, lists, and in the Saving Our Birds chapter. The scientific names of all other birds appear on page 188.

EUROPEAN BEE-EATER

BIRD ANATOMY

From the sharp end of the beak to the longest tail feather and outstretched wingtips, let's take a closer look at all the parts that make up a bird.

WINGS
The part of a bird's body used for flying

CROWN
Top of the head

NAPE
Back of the neck

THROAT

BREAST
Upper front of the body

SIDE

BELLY

MALLARD

RUMP
Lower back area

UPPERTAIL AND UNDERTAIL COVERT FEATHERS
Feathers that cover the top and bottom parts of the base of a bird's tail feathers. Covert feathers help keep airflow smooth over a bird's tail.

For most birds, wings are used for flying.

Bird wings are made up of different parts and different types of feathers that help in flight.

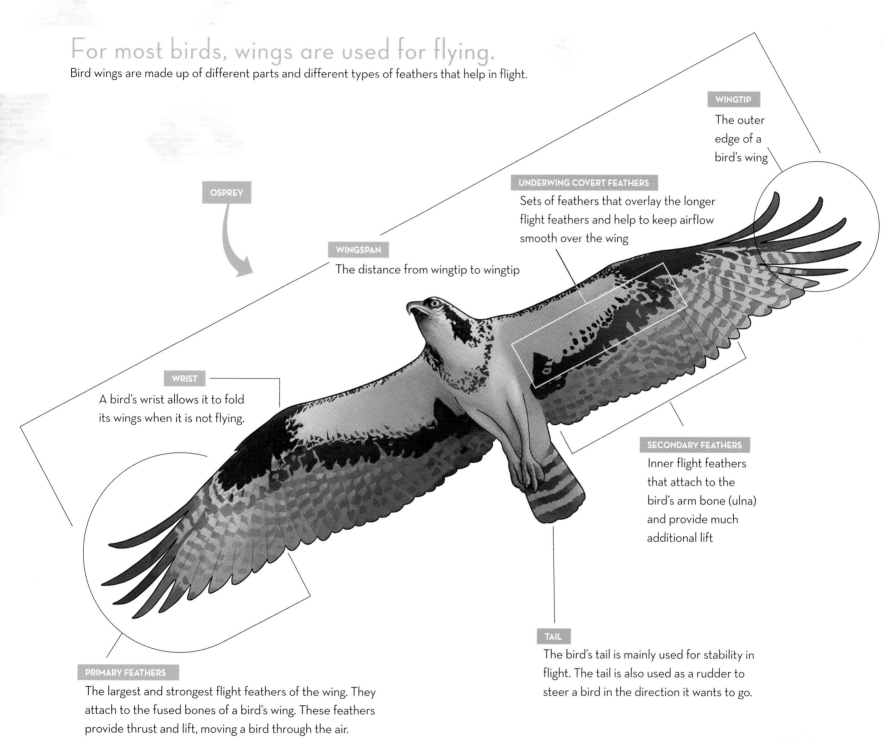

OSPREY

WINGTIP
The outer edge of a bird's wing

UNDERWING COVERT FEATHERS
Sets of feathers that overlay the longer flight feathers and help to keep airflow smooth over the wing

WINGSPAN
The distance from wingtip to wingtip

WRIST
A bird's wrist allows it to fold its wings when it is not flying.

SECONDARY FEATHERS
Inner flight feathers that attach to the bird's arm bone (ulna) and provide much additional lift

TAIL
The bird's tail is mainly used for stability in flight. The tail is also used as a rudder to steer a bird in the direction it wants to go.

PRIMARY FEATHERS
The largest and strongest flight feathers of the wing. They attach to the fused bones of a bird's wing. These feathers provide thrust and lift, moving a bird through the air.

11

# EGGS

Birds are not the only creatures that lay eggs.

Alligators and turtles lay eggs. Frogs and toads, too. Fish, snakes, cockroaches, crickets, praying mantises, duck-billed platypuses, and echidnas lay eggs. But birds' eggs have their own special qualities.

Birds' eggs are as varied as the birds themselves: round, ovoid, large, or small. Scientists tell us that the shape of the egg is determined by the membrane inside. As the chick in its little sac develops inside the egg, it puts pressure on the shell. Is it to become a wading bird with long legs? A good flier with a compact body? A chick with extremely broad wings? The growing bird's needs control the shape of the egg and shell.

Eggs also come in many different colors—white, brown, tan, gray, pink, blue, spotted, multicolored—they are almost gemlike in their variety. Those colors come from two basic pigments made in the bird's shell gland. One produces variations of reddish-brown colors, the other produces blue-green colors. But interestingly, scientific research has shown that the color is not produced until a few hours before the egg is laid.

As varied as bird eggs are, they all have several things in common. They all have a hard calcium carbonate shell. And they all have a yolk and egg white that contains what will grow—if fertilized—into a chick.

Those eggs are grown inside the hen, which is the name for any female bird. The hen deposits the eggs into a nest. Together, the eggs in the nest are called a clutch. Some birds, such as Laysan albatrosses, lay only one egg at a time. Red-tailed hawks lay two or three. And some birds lay a lot more. A mother wood duck lays between seven and fourteen eggs in her clutch. That's a lot of eggs and a lot of chicks!

ROBIN'S EGGS IN A NEST

Two eggs, one goose

BIRD EGGS IN A NEST

A ROBIN HATCHLING

LAYSAN ALBATROSS

The arithmetic of eggs
is hard to unscramble.
Laying, lying,
a tangle, a bramble.

Plus or minus,
one on one?
Zygotes, passion,
done, undone?

Two in the nest—
Quite round. Ovate?
Waiting their very
uncertain fate,

Before uncurling
beak or leg.
If there's a break-in . . .
Sum: goose egg.

—Jane Yolen

13

NESTS

Birds are warm-blooded creatures.

That means they need to keep their eggs warm in order for them to hatch. This is known as incubation. Birds sitting on eggs are vulnerable to predators, as are the eggs themselves. To protect their eggs, birds create a safe place for this incubation period. That place is a nest.

In most species, female birds are the nest builders and the male birds help. But there are some male birds that take on the nest-building task. One example is the male weaver. For these birds nest building is a courtship display that helps the weavers attract a female.

Many birds build their nests, but not all species do. Some birds just lay their eggs on rock ledges or bare soil or, in the case of cowbirds, in nests belonging to other birds. That means that another mother raises the cowbird chicks.

Other birds, such as penguins and flamingos, nest in colonies. A nest colony is made of multiple nests close together. Scientists believe some species nest this way for better protection against predators. Because there are many birds in a colony, they are better able to defend themselves.

Some birds make several nests in a season; some return to the same nest year after year. Some birds nest inside holes and some in the eaves or on top of man-made structures. Some birds roost (sleep) in their nests, and some only use nests during breeding season, when it's time to lay eggs and raise chicks. In other words, there is no one size or one way when it comes to nests and birds.

VITELLINE MASKED WEAVER MAKING A NEST

14

The Fallen Nest

I found it sulking on the hard ground
crowned by uncurling ferns,
wind-blown, two blue halves of a shell,
all that I could learn

of its lineage, its line of woven fabric
from the forest's store.
I took it home, gave it shelf,
could do nothing more.

I wondered if a robin in some storm
had felt the loosening of nest
falling away from her
accommodating breast.

Did she mourn the rest of the night,
Or simply beat her wings against the storm,
taking flight?

—Jane Yolen

HOME SWEET HOME

There are many different kinds of bird nests.

They can be simple or elaborate, high up in a tree or cliff, or nestled among shore-line reeds or right on a sandy beach. Some birds nest in colonies (such as terns); some nest alone (like ospreys). Nests come in as many shapes and sizes as there are types of birds.

BROWN-HEADED COWBIRD

Cowbird parents don't bother making nests. They lay their eggs in other birds' nests and let those foster parents raise their cowbird chicks.

BALTIMORE ORIOLE

The female Baltimore oriole spends a whole week weaving a hanging nest out of natural and man-made fibers. When finished, it is a deep pouch hanging from a branch.

BLACK KITE

These medium-size raptors line their nests with plastic. Scientists think they do this to show off to other kites.

16

BARN SWALLOW

Both the male and female barn swallow collect mud and grass stems in their bills to build their mud nests in man-made structures.

BALD EAGLE

These large birds of prey make enormous nests, called aeries, out of sticks. They return to the same nest every year and add to it. Eventually, an eagle nest can get so large that a human could sit in it. The biggest on record was 10 feet (3 m) wide and 20 feet (6 m) deep.

BURROWING OWL

Though they sometimes use abandoned holes dug by other animals, most burrowing owls dig their own holes in the ground for their nests. The holes are four to eight feet (1.2–2.4 m) long and, at the end, they open up into a cavity where the owls lay their eggs.

HAIRY WOODPECKER

Both the male and female hairy woodpecker share the job of making a nesting hole in a tree. Once completed, the hollowed-out nest is lined with the wood chips and sawdust created in the excavation, which cushions the eggs.

BLUE JAY

Male and female blue jays build a bowl-shaped nest in a tree using twigs, grasses, string, bark, and other, mostly natural, materials.

BEAKS

All birds have beaks, but not all beaks are created equal.

Beaks, both the upper and lower mandibles, are made out of keratin, the same material as human fingernails and horns and hooves in other animals. Birds are not the only creatures with beaks, though. Some insects have beaks. Turtles, cuttlefish, octopuses, and squid have beaks. So does the duckbilled platypus.

A bird's beak is used for many different things, including preening, house building, feeding chicks, and defense. But the most important job for every bird's beak is gathering and eating food. Beaks are specialized for the type of food each bird eats.

So what do birds eat? It depends on the type of bird. Many birds eat seeds and nuts and berries. Some birds eat farm crops, such as corn, rice, and wheat. Raptors (hawks and their cousins) eat meat. Owls eat mice, rats, shrews, voles, and the occasional weasel. Some birds actually eat other birds, and also eat other birds' nestlings and eggs. Ravens, crows, and vultures are great street cleaners. They eat dead animals, called "carrion," that have been killed on our busy highways. Birds drink water, but some species drink very little, getting most of their water from the food they eat.

Shorebirds eat fish, water plants, even crustaceans. Some eat sea stars. Robins eat worms. Many birds eat insects. And hummingbirds sip nectar from flowers. All birds are specially adapted for their food and water needs.

A MALE BIRD OF PARADISE KNOWN AS THE PARADISE RIFLEBIRD

FLAMINGO

BROWN PELICAN

A Beak

A seed,
a sip,
a stump,
a drip.

To crack,
to rake,
to peck,
to break.

A sword,
a meal,
to fight,
to steal.

To serve,
to call—
a beak
for all.

—Heidi E. Y. Stemple

VON DER DECKEN'S HORNBILL

BEAUTIFUL BEAKS

There are many different kinds of beaks (also called bills).

Bird beaks are adapted to the type of food a bird eats. Beaks need to be able to get at the food, crack it open, take out the inedible parts, and then get the food either to the bird itself or its chicks. They can be long or short, colorful or plain, hooked or straight.

OLD WORLD WARBLERS

These insect-eaters have small pointed beaks ideal for capturing all sorts of bugs.

DUCKS

Soft beaks help these aquatic birds "feel" their food. Just inside the beak are comblike "teeth" (called lamellae) that help to filter out water and mud from their favorite foods, which include insects, small crustaceans, grasses, and fish.

PENGUINS

Long and sharp, penguins' beaks are perfect for chomping prey underwater.

PELICANS

Fish-eating pelicans can capture a mouthful of dinner with their pouch-like beaks, which are used to scoop up food and filter out water.

WOODPECKERS

These omnivores have strong beaks with chisel-like tips that can hammer and drill into wood. Once inside, their extra-long barbed tongue, covered with sticky saliva, grabs delicious insects.

SPOONBILLS

Waving their long spoon-like bills back and forth in shallow water, spoonbills feel for their food. When they feel a bug or a little fish, chomp! They eat up a tasty meal.

BONES

In 1683, the Italian astronomer, physicist, and philosopher Galileo wrote that bird bones were light-weight. It turns out that isn't true.

If you hold a songbird in your hand, it will seem to weigh almost nothing. But actually, bird skeletons weigh about the same as the skeletons of mammals of similar size. The bones of birds, however, are denser, stronger, and more rigid than other creatures' bones and, in many places, they are fused together. All of these adaptations make flight possible. Wing bones certainly can't afford to be fragile. They have to withstand the pressure of the twists and turns of the wind on takeoffs, long flights, and landings.

Another bird bone adaptation is a keeled (shaped like a ship's keel) sternum, also known as a breastbone. This is where flight muscles are attached. The wishbone, the central Y-, V-, or U-shaped bone in a bird's chest, helps stabilize the chest cavity for flight. Birds also have more neck (cervical) vertebrae than many other animals. Humans have seven neck vertebrae. But most birds have 13 to 25 of these very flexible neck bones. Why? Well, for one thing, it helps them groom their feathers.

Some bird bones are hollow. Instead of being filled with marrow like most other creatures' bones, hollow bones have an air cavity inside. In some birds, these have become extensions of lung air sacs, which can fill up with extra oxygen for long flights.

That strong sternum and those hollow-boned wings carry a vee of geese across the sky and allow the arctic tern to fly 44,000 miles (70,800 km) on its migration. These features also protect the osprey in its headlong dive into the water from a great height to catch a tasty fish dinner.

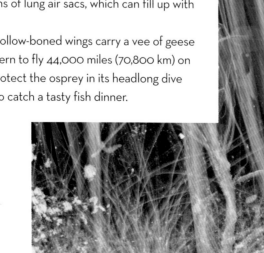

AN X-RAY OF A WADING BIRD KNOWN AS A SNIPE. SNIPES ARE IN THE SAME FAMILY AS OTHER SHOREBIRDS, SUCH AS SANDPIPERS AND TURNSTONES.

Hollow Bone

Does the bird play the hollow
of its long wing bone,
When it's off in the woods,
Still mate-less and alone?

Does it let the wind whistle
through the hollow in its breast,
Does it sing a simple lullaby
While flying to its nest?

Does the keel of its sternum
Vibrate with earnest tune?
Does the goose head for home
Singing carols to the moon?

—Jane Yolen

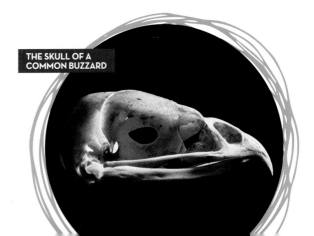

THE SKULL OF A COMMON BUZZARD

WINGS

"You cannot fly like an eagle with the wings of a wren."
—William Henry Hudson

Most birds have wings. In fact, each species of bird has wings adapted specifically for that bird. Some are short and pointed wings (hummingbirds). Some look like long fingers (crows). Some have feathers so perfectly shaped that when the bird flies it makes no sound at all (owls).

But birds are not the only winged creatures on Earth. Bats have wings. Insects have wings. There are flying lizards; some of these lizards are winged geckos. There is even a flying squirrel (though really it just glides). Wings are for flying, but in fact not all birds can fly. There are over 60 species of birds now alive that cannot fly, emus and penguins among them. And some birds—like kiwis and cassowaries—don't have any wings at all.

The smallest flightless bird in the world has the best name: the Inaccessible Island rail. It is found in only one place, an extinct volcano in the South Atlantic Ocean called Inaccessible Island, which is, in fact, quite difficult to reach.

The largest flightless bird is the ostrich, which uses its wings only for balance while running. The ostrich may have weak wings, but look out for its feet because it can deliver a strong and dangerous kick.

For most birds, though, wings are used for flying.

GREAT KISKADEE

24

To Fly or Not to Fly

To fly or not to fly?
If that is the question,
Evolution is the answer.

The penguin
knows nothing of the sky.
He traded his aerial skills
for the depths of the
Antarctic sea.

The ostrich
does not care about
elevated travel.
Her wings are for show only.
And her feet
can cover many miles.

The flightless cormorants,
like their large-winged cousins,
still dry appendages
in the sun.
As if readying them
for the flight they will never take.

To fly or not to fly?
Evolution chooses the wing
and the path—
the sea, the land, the sky.

—Heidi E. Y. Stemple

A hummingbird can beat its wings up to 80 times a second.

NORTH ISLAND BROWN KIWI

25

WONDROUS WINGS

Not all birds use their wings for flying.
But those that do have wings specialized for specific types of flight.

PASSIVE SOARING WINGS
Made for catching the wind, these wings are made up of long primary feathers that have slots at the end and look almost fingerlike.

BALD EAGLE

ACTIVE SOARING WINGS
Built for long-distance soaring, these wings are long and narrow.

EURASIAN TREE SPARROW

ELLIPTICAL WINGS
These wings are good for fast, easy maneuvering in tight spots, but only for short periods of time.

WHITE TERN

HIGH-SPEED WINGS
Birds that need to fly fast for long periods of time have these long, thin wings.

WANDERING ALBATROSS

HOVERING WINGS
These small, fast-moving wings make a bird look like it's floating.

WHITE-BELLIED MOUNTAIN GEM

WINGSPANS

A wingspan is the distance from the tip of the outermost feather of one wing to that of the other.

Wandering albatross
Diomedea exulans
11.5 feet (3.5 m)

California condor
Gymnogyps californianus
10 feet (3 m)

Magnificent frigatebird
Fregata magnificens
7 feet (2.1 m)

Ostrich
Struthio camelus
6.5 feet (2 m)

Mallard
Anas platyrhynchos
3 feet (0.9 m)

Western screech-owl
Megascops kennicottii
2 feet (0.6 m)

European starling
Sturnus vulgaris
15 inches (38 cm)

Cactus wren
Campylorhynchus brunneicapillus
11 inches (28 cm)

Calliope hummingbird
Selasphorus calliope
4 inches (10 cm)

Bee hummingbird
Mellisuga helenae
less than 3.25 inches (8.3 cm)

27

FEATHERS

There's a Yiddish phrase: A bird is known by its feathers.

Of all living creatures, only birds have feathers.

Long feathers, short feathers, straight feathers, curved feathers, downy feathers, pointed feathers, brightly colored feathers, black or white feathers. All kinds of feathers. And each feather has its own job. Feathers attach to the bird by the quill (which is the bottom part of the shaft of the feather) and help with flight and swimming, temperature control, and attracting a mate.

Paleontologists point out that a good many dinosaurs probably had feathers, too. Scientists are discovering more and more dinosaur fossils in which the imprint of feathers can be seen. So scientists now believe that dinosaurs were the ancestors of birds.

That means if you want a short, accurate definition of a bird, simply call it the living thing with feathers. And you will be right.

RED-TAILED BLACK-COCKATOO

Shake a Tail Feather

FLAMINGO FEATHERS

For keeping warm
and in the air,
for camouflage
or flashy flair.

Attract a mate,
fly soundlessly,
lining nests,
or buoyancy.

FEATHERS OF THE BLUE-
FRONTED AMAZON PARROT

Weather-safe
and water-proof,
fit in, show off,
aloft, aloof.

Tropics, Arctic,
farm, backyard,
Feathers—
the avian
calling card.

—Heidi E. Y. Stemple

A PEACOCK SHOWING
ITS TAIL FEATHERS

29

PARTS OF A FEATHER

Each feather is made up of two main parts: the shaft and the vane. The shaft and vane are then made up of more parts. Here are the parts of a feather.

SHAFT

The central support for the feather, which runs from the base to the tip of the feather. It is made up of the rachis—the "stem" of the feather to which the vanes are attached—and the quill (or calamus), which is the part that attaches to the bird under the skin.

QUILL

VANE

The flattened part on both sides of the shaft. The vane of a feather is made up of one to three parts, depending on the type:

A. Barbs: If you think of a feather as a tree, these are the branches. They extend out from the shaft at a 45-degree angle.

B. Barbules: thin fibers that extend out from the barb

C. Barbicels: tiny hooks that reach out and interlock the barbules together, zipping the entire feather tightly

BARBS

RACHIS

BARBULES

BARBICELS

Feathers aren't all over a bird in random spots; they grow in rows.

Birds generally have between 6 and 32 tail feathers.

Many birds' feathers weigh more than their bones.

Woodpeckers have two stiff tail feathers that help prop them upright when clinging to a tree trunk.

The pink color of flamingos' feathers is caused by pink pigments in the shrimp they eat.

FABULOUS FEATHERS

Birds have seven types of feathers; each has its own form and function.

From the fluffiest down feather to body-covering contour feathers, each feather is specifically designed for its own job. They are important for insulation and flight, but also for camouflage. And feathers come in handy for displays of courtship and aggression.

BLUE-AND-YELLOW MACAW

WING
Wing feathers are asymmetric, meaning they are larger on one side of the shaft than the other. These feathers have interlocking barbs, which make them windproof. They are anchored to the bone, instead of just the skin like other feathers, and are used to help the bird generate thrust and lift for flight.

BRISTLE
Almost entirely shaft, these feathers (which don't really look like feathers) are found mostly on the head of the bird, around the mouth, eyes, and nostrils. Experts think these feathers are associated with a bird's sense of touch.

GREAT BUSTARD

TAIL
Tail feathers are generally similar to wing feathers in that they have interlocking barbs. The two central feathers are somewhat symmetrical and attached only to the skin. Most are used for steering during flight, but some are just ornamental.

LONG-TAILED WIDOWBIRD

32

SEMIPLUME
These fluffy feathers have no barbules or barbicels. They are close to the body and serve as insulation.

WANDERING ALBATROSS

FILOPLUME
These long, hairlike feathers have barbs only at the tip. Experts think they may have a sensory function.

GREAT CORMORANT

A CONTOUR FEATHER OF A MACAW

CONTOUR
Contour feathers cover the body of the bird and the wings where the wing feathers attach to the wing bones. These feathers form the outline of a bird's plumage and are half downy (near the body) and half waterproof (at the tip). The downy parts lack the barbules and barbicels that hold the barbs together. Contour feathers help to streamline a bird in flight.

A YOUNG MUTE SWAN

DOWN
These are the fluffiest feathers. They have no barbules or barbicels and lie close to the body. Down feathers provide most of the insulation for a bird, keeping it warm.

33

EPIDEXIPTERYX

THE ANCIENT HISTORY of BIRDS

Dinosaurs are the ancestors of modern birds.

ARCHAEOPTERYX

DINO BIRDS

Millions of years ago dinosaurs walked the Earth. They were huge lizardlike creatures that roamed the land, from the polar regions to everywhere in between. And then they went extinct. But what dinosaurs left behind were lots of fossils: imprints of their bones, skin, footprints, and sometimes fossilized eggs. These fossils are clues to how dinosaurs lived and how they went extinct. Paleontologists use dinosaur fossils to help unravel the mystery of those huge early creatures.

Most scientists believe that dinosaur extinction happened due to global climate change—going from the warm, mild Mesozoic era to a Cenozoic era that had many more changes in temperature. Where the scientists differ is in how quickly that change occurred. Some think it was gradual, others believe it may have been an outside disaster—like a major asteroid strike. But they all agree that the big dinosaurs suddenly disappeared around the same time.

Whenever new fossils are found, questions are answered, and different questions arise.

What did the scientists make of the earliest collected fossils? They had learned that not all dinosaurs resembled lizards. Some seem to have been birdlike, with keel-shaped breastbones, long beaks, and hollow bones.

There were lots of arguments.

And then, in the 19th century, a paleontologist in Germany discovered fossil evidence of feathers.

Feathers? The fossil evidence was very clear.

A *T. rex* with feathers? Some people scoffed at the idea that dinosaurs could have been the first birds.

BUT THE EVIDENCE GREW. AND GREW. AND GREW.

ARCHAEOPTERYX

This is how those discoveries went.

ARCHAEOPTERYX

For a long while, scientists considered *Archaeopteryx* (ARK-ee-OP-turr-icks), the "first bird." Its fossil was the initial spark for the notion that birds are the living remnants of the dinosaur line. But as the years went by, *Archaeopteryx* remained the single piece of fossil evidence that there was a line of dinosaurs that were actually early birds.

The story of *Archaeopteryx* began in 1861, when workers in a German limestone quarry discovered the impression of a 145-million-year-old feather.

Paleontologist Hermann von Meyer studied the feather impression. At first he thought it was a clever forgery. But when he compared the impressions of the feather on the upper and lower layers of limestone in which it was found, he declared, "No draughtsman could produce anything so real."

Von Meyer gave the fossils the name *Archaeopteryx*, meaning "ancient wing." Later in 1861, the first known skeleton of *Archaeopteryx* was discovered in Germany. It was sold to the Natural History Museum in London, where in 1863, paleontologist Richard Owen described it as a bird.

The idea of *Archaeopteryx* as the "first bird" even intrigued Charles Darwin, who had published *On the Origin of Species* just three years earlier. Reportedly, he said to a friend, "It is a grand case for me," meaning it supported his theory of evolution.

One hundred and fifty years later, in the late 1900s, things started to get more complicated. New fossils, which were being dug up at a rapid pace, revealed a gradual transformation from ground-dwelling dinosaurs to birds capable of flight. These newly discovered fossils were older

XIAOTINGIA ZHENGI

than *Archaeopteryx*, making it clear it was not the first bird. The fossils not only had imprints of feathers, they also had wings with three digits and a wishbone in the body. In other words, they had even more features in common with modern birds.

XIAOTINGIA ZHENGI

The next candidate for first bird was *Xiaotingia zhengi* (shyow-TIN-gee-uh zeng-ghee). This fossil was found in western Liaoning, China, in rocks dated between 161 million and 145 million years old.

It was named after Zheng Xiaoting, a scientist who helped establish the Shandong Tianyu Natural Museum for Chinese vertebrate fossils.

The rock impression showed a creature with claws on the ends of its forelimbs and sharp teeth. Also, *Xiaotingia zhengi* had a wishbone. But soon, *Xiaotingia zhengi*, too, was overshadowed by earlier fossils with more birdlike features.

EPIDEXIPTERYX

In 2008, analysis by bird and dinosaur specialists suggested that the earliest known avian was a pigeon-size, feathered creature they called *Epidexipteryx* (Ep-pid-decks-XIP-terricks). This middle-to-late Jurassic fossil from 168 to 152 million years ago was discovered in Inner Mongolia, China. But more fossils were to come.

ARCHAEORNITHURA MEEMANNAE

In 2015 in the wetlands of China, two new fossils were discovered. They were named *Archaeornithura meemannae* (ARE-kee-or-neth-THUR-a MEE-man-eye). *Archaeornithura* means "ancient ornithuromorph," and *meemannae* was chosen in honor of the female Chinese paleontologist Meemann Chang.

ARCHAEOPTERYX

ARCHAEORNITHURA MEEMANNAE

Each specimen of this dinosaur had the telltale traits of a modern bird: fan-shaped tail feathers, highly fused bones at the ends of the wings, and a U-shaped wishbone.

One more bird feature clinched them the title of earliest birds: a small projection on the front edge of their wings. This was important because it is a feature similar to that of modern birds. It helps boost a bird's ability to maneuver in the air.

Another important discovery was that *Archaeornithura* had long legs and feet that seemed adapted to wading in water.

All these new finds prompted Chinese paleontologists Min Wang and Zhonghe Zhou to suggest that dino birds had already developed the ability to fly by the time of *Archaeornithura meemannae*, 130 million years ago.

TONGTIANLONG LIMOSUS

A new fossil discovery in 2016 shed even more light on bird evolution. A dinosaur with a beak and feathers was studied carefully in the Ganzhou region of China. It had been unearthed four years earlier during the construction of a school. Paleontologists called it *Tongtianlong limosus* (Tong-shin-long lim-O-sus), which means "muddy dragon on the road to heaven." It was the size of a donkey. Although it clearly had a beak and feathers, it didn't otherwise have the more sophisticated tail or wishbone. It was not a true bird—but what was it? The scientists suggested it was an oviraptorosaur. This dinosaur shares a common ancestor with the lineage that later evolved into birds.

This discovery raised more questions about how birds evolved. Could it be that not all dinosaurs with feathers were birds—that they were some other kind of creature? Does it mean they couldn't fly? Does it mean they wouldn't have developed into flying birds, but rather into birds like ostriches and emus, which can't get off the ground?

Each new fossil discovery means that scientists have to regroup and create new definitions. It means refining what we know and what we don't know about birds. And it means scientists will continue to make new discoveries.

Once A Week from the Paleontologist

Once a week
new dinosaur bones
emerge from catacombs,
earthy tombs.

Once a week,
we see them fly,
from the last outposts.
We posit sky.

Call them birds,
remark the sight,
create a Jurassic
of endless flight.

—Jane Yolen

TONGTIANLONG LIMOSUS

GOLDEN EAGLE

BIRDS in HISTORY

Birds have lived alongside humans throughout history.
Sometimes their stories intersect.

DOMESTIC GEESE ON THE RUN

MAN'S BEST FRIEND

Even though the job is named after them, the best watchdogs in the world might not be dogs at all.

The best guard a person could have might just be a domestic goose. Or a flock of geese.

Geese are used to guard all kinds of things all around the world: from a whiskey distillery in Scotland to a prison in Brazil, from police stations in China to military installations in Germany. In the United States, farmers often use geese to guard their chickens. The big birds will tangle with predators the size of coyotes to protect their smaller cousins.

The first known example of geese protecting people was in ancient Rome around the year 390 b.c. A group of people known as the Gauls tried to invade Capitoline Hill to kill the last Romans defending the city. But a flock of geese raised such a racket that the defenders awoke and fought off the invaders. Dogs had been set out to guard the Romans' camp that night. But they hadn't made a peep!

What makes geese such good watchdogs? Aside from being incredibly loud, geese are extremely territorial. Even though he could fly away from danger, when a male goose's flock is threatened, he will fight. A male goose has a serrated beak and wings that can break bones when he buffets his enemies with them. And though wild geese only grow to around eight pounds (3.6 kg), domestic geese, including the geese kept by the Romans, can top 20 pounds (9.1 kg).

That's a lot of goose to fight.

Add to that great hearing, great eyesight, and the fact that they can be fed on mostly grass, and you have one of the world's best watchdogs in a big, beaked, feathery package.

A MONGOLIAN CHILD
HOLDING A GOLDEN EAGLE

FALCONRY IN THE MIDDLE AGES AND BEYOND

Falconry is a sport in which a bird of prey (known as a raptor) is trained to hunt and bring back prey to a person's hand. In exchange, the raptor gets a reward, usually a piece of meat.

Falconry has been around for centuries. A statue from seventh-century b.c. Assyria (now part of Iraq, Turkey, and Syria) is probably the oldest known reference to the sport. There were also references to falconry in China as early as 680 b.c., and a mention in a Japanese source that falcons were given as hunting gifts to Chinese princes in 220–206 b.c.

The interest in British and European falconry was stimulated by trade between Arabia, Europe, and the Far East. But falconry was not popular in the West until it reached the Mediterranean countries in 400 a.d. By 875 a.d., falconry was practiced widely through western Europe and England. In fact, its popularity may have been one of the reasons for the rise of a new science: ornithology—the scientific study of birds. The Holy Roman Emperor Frederick II of Hohenstaufen was a fanatic falconer. He wrote a book about falconry that took him over 30 years to finish: *De Arte Venandi cum Avibus* (The Art of Falconry). It was one of the earliest scientific books on the anatomy of birds.

Perhaps the one place in the world where true hunting (not sport) with birds remains is on the western plains of Mongolia. There a nomadic group of Kazakhs continue an ancient tradition. They hunt with golden eagles. Each hunter carries an eagle on an outstretched arm while riding a pony across miles of rocky landscape. The nomads hunt for food, but they also hunt foxes for their pelts. At the end of seven years, the Kazakhi hunter sets the trained eagle free.

An even more modern twist to this practice takes place in France. In 2016 and 2017 the French military began training four golden eagles to take down drones in no-fly zones. Why were golden eagles chosen? Like all eagles, golden eagles haven excellent eyesight. They can spot prey from a mile (1.6 km) away. Their talons are also powerful and can exert a pressure of 500 pounds per square inch (35 kg per sq cm). This is at least 10 times stronger than that of an adult human's hand. The eagles can also swoop downward at speeds of over 150 miles an hour (242 km/h). With this kind of speed and strength, the four trained eagles can easily catch drones.

The four eagles in the program are named after the famous musketeers: Aramis, Athis, Porthos, and D'Artagnan. The soldiers train the eagles from the time they are chicks, using drones as dinner plates so that the birds always associate drones with food. As the eagles grow older, the trainers bring the birds outside, where drones hover in the air. The eagles seize the drones. Once the drones are seized, the eagles fly to the ground with their prey and "mantle" over, or cover, the mechanical prey with their wings. For their efforts, the eagles are rewarded with a piece of meat.

RAVENS OF THE TOWER

"If the ravens leave the Tower, the kingdom will fall." Nobody knows exactly when this legend began, nor when the ravens arrived. But as long as the people of England can remember, they've believed that if the ravens left the Tower of London, something terrible would befall their beloved island.

In the middle of the 17th century, there was a royal observatory in the Tower of London. The royal astronomer worked his craft there, investigating the stars and the skies, and reporting his findings to the king, Charles II. There was just one problem.

Ravens.

Wild ravens made the tower their home. They flew in front of the lens of the astronomer's telescope when he tried to make observations. They stole food and shiny things from his laboratory. They left their droppings on his telescopes. It had to be stopped.

Luckily for the royal astronomer, King Charles II liked hearing about the stars and the skies, and didn't much like hearing about stolen lunches and soiled lenses. So he decreed that the ravens be destroyed, and nevermore darken his royal astronomer's lenses with their feathers or their droppings.

But before the hunters could load their weapons or the poisoners lay their traps, the king was visited by an old seer. The seer told him of an ancient legend that declared that the tower must always have at least six ravens living in it or England would fall.

The king was a sensible man. And sensible men in the 17th century didn't risk the destruction of their kingdom over one disgruntled astronomer's feud with a bunch of birds. King Charles II changed his mind and decreed that there would always be at least six ravens living in the tower, and thus the safety of England was preserved.

Only ...

If you look for the decree, you won't find it.

If you look for ravens in the listing of the animals in the Royal Menagerie kept at the tower since the 13th century, you won't find them.

If you look for any mention of ravens at the tower before their 1883 debut in a handout for the tower called "The Pictorial World," once again, you won't find anything.

And yet ...

Ravens are native to Britain and have lived there for thousands of years. They are a sturdy and adaptable bird, well suited to city life. They would have been common in London at the time and would definitely have been seen regularly at the tower. Why no mention of them? It's a mystery.

The tower ravens of legend are now the tower ravens of fact. Because no matter their origin, they are there in the tower. There are always at least six of them. The ravens' wings are clipped so they won't fly away. They are even enlisted in the army. They are cared for and bred and kept safe. Because, like the king long ago, legend or not, the caretakers of the tower aren't going to risk the fate of their country over these six birds.

COMMON RAVEN

AUDUBON CHRISTMAS BIRD COUNT

It was Christmas in the year 1900. The game hunters were getting ready for their annual tradition of splitting into teams and setting out on "side hunts" to see how many birds they could count. And by "count," they meant shoot. But Frank Chapman, an ornithologist who worked at New York City's American Museum of Natural History, published an announcement in his magazine, *Bird-Lore*. Instead of counting dead birds, he suggested, let's count them with our ears and eyes. Let's count them and keep them alive.

Twenty-seven bird-watchers in 25 locations did just that. It was the very first Audubon Christmas Bird Count. *Bird-Lore* magazine proudly printed the results—close to 18,500 individual birds from 89 different species were counted.

In 2014, which marked the 114th count, 71,659 birders across the Western Hemisphere participated in 2,408 count circles. They counted 66,243,371 individual birds of 2,403 different species.

The Audubon Christmas Bird Count (CBC) is a citizen science project. That means citizens—any person—can act as a scientist. Local birding clubs and Audubon chapters are assigned a specific area. For 24 hours, they count birds and compile data to report to the National Audubon Society. This data helps determine which bird species are in trouble or declining, and how climate change is affecting birds. The data collected has been used by the U.S. Environmental Protection Agency (EPA) as one of the 26 indicators of climate change.

For more information about the Audubon Christmas Bird Count, including how to find and join the CBC or other counts, please see the Find Out More section on page 184.

DURING THE CHRISTMAS BIRD COUNT, BIRDERS COUNT EVERY BIRD THEY SEE IN THE NAME OF CITIZEN SCIENCE.

BOHEMIAN WAXWING

THE PEABODY DUCKS

The Peabody Hotel in Memphis, Tennessee, U.S.A., has a long and storied history dating back to 1869, but it is best known for a quirky tradition. Every day for nearly 80 years, five ducks have marched through the swanky hotel lobby to go splash around in the marble fountain.

The first feathered fowl to enjoy the fountain were a group of English call ducks. Call ducks are domesticated ducks that were originally used as live decoys for duck hunting (a practice that is now illegal in most places). Decoy ducks would be put in a pond where they would attract wild ducks that hunters, hidden nearby, could shoot.

The English call ducks had just returned from a hunting trip to Arkansas in the early 1930s with the hotel's general manager. He and his hunting buddy thought it would be funny to let the ducks swim in the fountain. The next morning, hotel guests were delighted at the frolicking ducks and an unlikely tradition was born.

The ducks turned out to be such a hit that in 1940, the hotel bellman, Edward Pembroke, became the official Peabody Duckmaster. He used his skills as a former circus trainer to teach the ducks to march daily to and from the fountain. Mr. Pembroke continued the tradition for 50 years, each day marching the ducks from their rooftop home, down the elevator, and through crowds of adoring fans along the red carpet runway for their daily splash.

After the original English call ducks retired, mallards were brought in to take their place. The first group of mallards—one drake and four hens—assumed their title as the Peabody Ducks for three months before retiring so that a new batch could enjoy the limelight.

The tradition continues today and with a 2014 renovation to the ducks' penthouse residence, the pomp and circumstance of their daily march seems likely to continue into the future.

Counting Birds

1 bright red cardinal
at the top of a tree,
2 tufted titmice
defying gravity,
3 piping plovers
playing in the surf,
4 noisy fish crows
defending their turf,
5 rufous hummingbirds
sip nectar from flowers,
6 diving least terns
I could watch them for hours,
7 turkey vultures
cleaning up at roadside,
8 mallard ducklings
keep their parents occupied,
9 red-tailed hawks
riding thermals overhead,
10 different raptors
hunting in the watershed.
Just one more bird, I promise ...
it's so hard to resist.
Each one's a check mark
on my birding life list.

—Heidi E. Y. Stemple

DUCKS AT THE PEABODY HOTEL IN ORLANDO, FLORIDA, U.S.A.
VISITORS AT THE MEMPHIS HOTEL LOVED THE FOUNTAIN DUCKS
SO MUCH, IT WAS MADE A TRADITION AT OTHER PEABODY HOTELS.

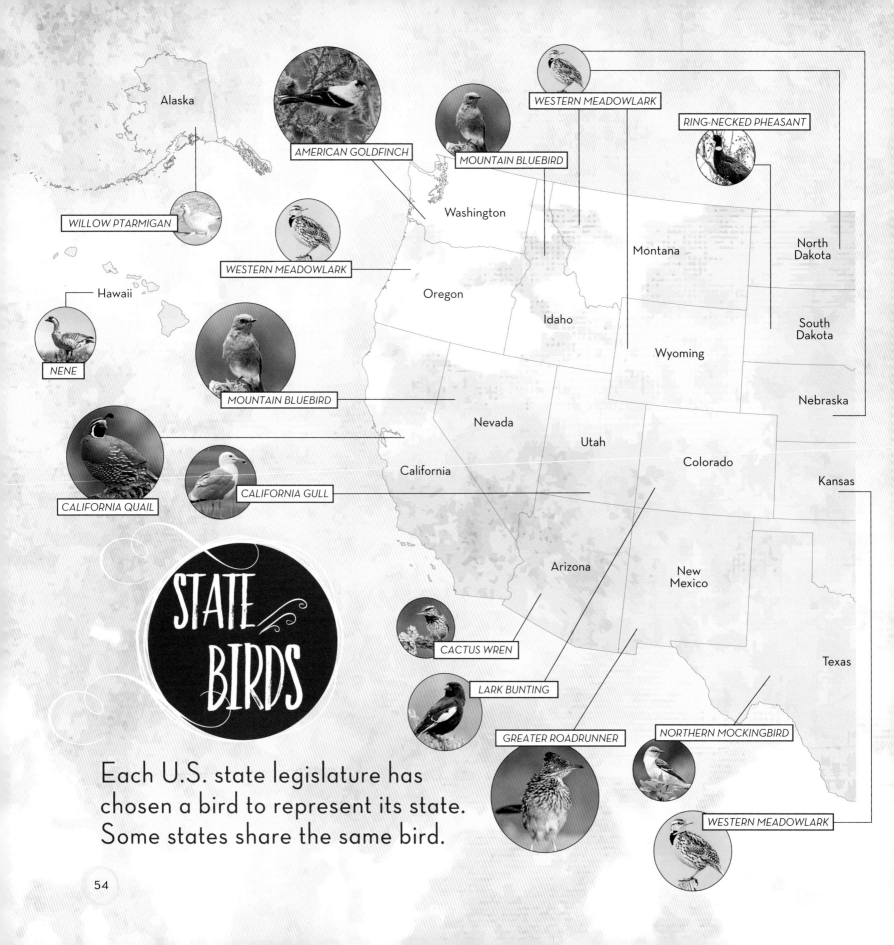

STATE BIRDS

Each U.S. state legislature has chosen a bird to represent its state. Some states share the same bird.

WILLOW PTARMIGAN

AMERICAN GOLDFINCH

MOUNTAIN BLUEBIRD

WESTERN MEADOWLARK

RING-NECKED PHEASANT

NENE

WESTERN MEADOWLARK

MOUNTAIN BLUEBIRD

CALIFORNIA QUAIL

CALIFORNIA GULL

CACTUS WREN

LARK BUNTING

GREATER ROADRUNNER

NORTHERN MOCKINGBIRD

WESTERN MEADOWLARK

Alaska

Hawaii

Washington

Oregon

Idaho

Montana

Wyoming

North Dakota

South Dakota

Nebraska

Nevada

Utah

Colorado

Kansas

California

Arizona

New Mexico

Texas

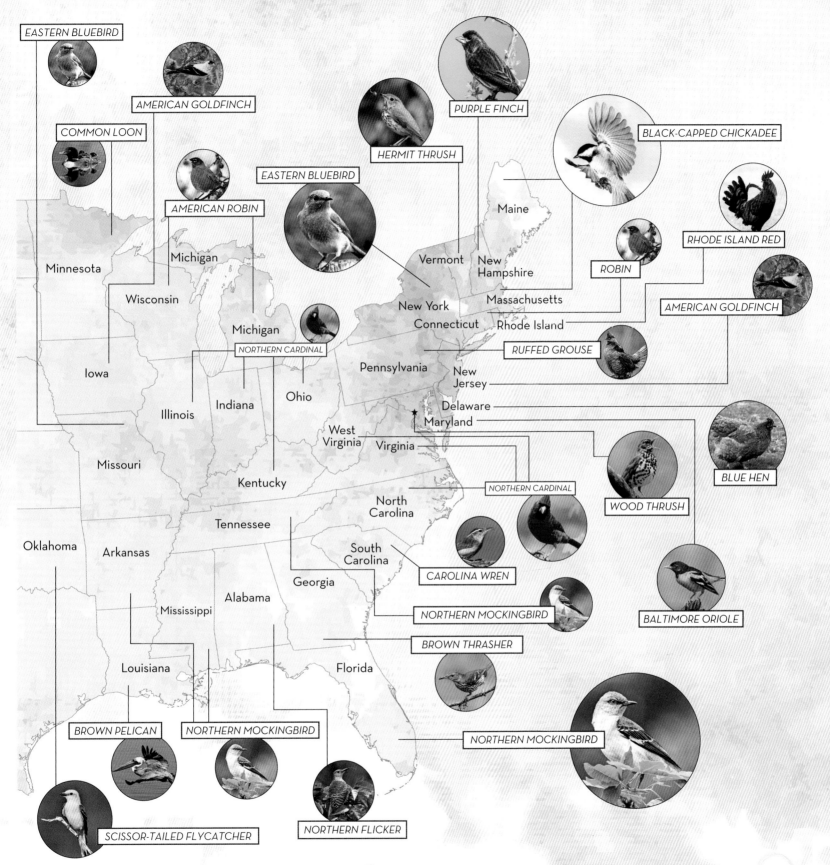

EASTERN BLUEBIRD

AMERICAN GOLDFINCH

COMMON LOON

AMERICAN ROBIN

EASTERN BLUEBIRD

HERMIT THRUSH

PURPLE FINCH

BLACK-CAPPED CHICKADEE

RHODE ISLAND RED

ROBIN

AMERICAN GOLDFINCH

NORTHERN CARDINAL

RUFFED GROUSE

Minnesota

Michigan

Wisconsin

Michigan

Iowa

Illinois

Indiana

Ohio

Missouri

West Virginia

Virginia

Kentucky

BLUE HEN

WOOD THRUSH

NORTHERN CARDINAL

North Carolina

Oklahoma

Arkansas

Tennessee

South Carolina

CAROLINA WREN

Georgia

BALTIMORE ORIOLE

Alabama

Mississippi

NORTHERN MOCKINGBIRD

Louisiana

Florida

BROWN THRASHER

BROWN PELICAN

NORTHERN MOCKINGBIRD

NORTHERN MOCKINGBIRD

SCISSOR-TAILED FLYCATCHER

NORTHERN FLICKER

Maine

Vermont

New Hampshire

New York

Massachusetts

Connecticut

Rhode Island

Pennsylvania

New Jersey

Delaware

Maryland

55

NORTHERN FLICKER

Colaptes auratus

The northern flicker digs up ants with its curved bill. The bird then uses its long barbed tongue to capture the ants and pull them into its mouth.

The northern flicker is sometimes called the yellowhammer.

56

This chicken-like bird sounds as if it is laughing, with its jumble of clucks, barks, rattles, and chortles.

These birds often fly into snowbanks to sleep, leaving no tracks for predators to follow.

Alaska

WILLOW PTARMIGAN

Lagopus lagopus

CACTUS WREN

Campylorhynchus brunneicapillus

Cactus wrens get almost all their fluids from the insects and fruit they eat.

If a predator gets too close to a cactus wren's nest, the birds will create a mob to get rid of the intruder.

58

Throughout a mockingbird's life, it learns more and more songs. Some individuals sing as many as 200 songs.

The mockingbird often sings into the night, especially if there is a full moon.

Arkansas, Florida, Mississippi, Tennessee, Texas

NORTHERN MOCKINGBIRD

Mimus polyglottos

CALIFORNIA QUAIL

Callipepla californica

Female quail sometimes lay their eggs in another quail's nest, which is called egg dumping. This means nests can have lots of eggs—almost 30 in some cases.

The California quail has a fancy plume on the top of its head that is made up of six feathers.

Lark buntings are actually a kind of sparrow. They build ground nests made of grass, roots, and stems.

Male lark buntings change their plumage with the change of the seasons. In summer they show their breeding colors: a distinctive black with white wings. When winter comes, the males change color to a drab, streaky brown-gray. The female stays the same color all year.

Colorado

LARK BUNTING

Calamospiza melanocorys

AMERICAN ROBIN

Turdus migratorius

Robins' eggs are thought to be blue to help protect them from light. Also, the bluer the egg, the more the father feeds the babies once they hatch.

Many robins stay in the same area all year long. In the winter months, robins spend less time on the ground and more time roosting in trees, so they aren't as easy to spot.

62

The blue hen is not only Delaware's state bird, it is also one of only two domesticated birds to represent a state.

Delaware

BLUE HEN

Gallus gallus domesticus

The blue hen was domesticated from a wild chicken known as the red junglefowl. Red junglefowl are native to southern Asia.

Georgia

BROWN THRASHER

Toxostoma rufum

Brown thrashers have quite a musical repertoire. They can sing more than 1,000 different songs.

When looking for food, brown thrashers thrash their beaks back and forth through the leaves, earning them their name.

The plumage of male and female nenes is identical, so it's hard to tell them apart.

Because of its habitat, the nene has evolved feet that have less webbing and are more clawlike for walking on rough Hawaiian lava rocks.

Hawaii

NENE

Branta sandvicensis

MOUNTAIN BLUEBIRD

Sialia currucoides

Most birds count on their plumage to attract females, but a female mountain bluebird largely ignores the stunning blue color of the males. Instead, she chooses a mate based on the quality and location of the nest he offers her.

Mountain bluebirds nest in cavities—natural hollows or woodpecker holes in trees, holes in a dirt bank or cliff, even holes in the side of a building.

66

The cardinal is known for aggressively attacking its own reflection in any reflecting surface.

The male cardinal is one of the most easily spotted and identified birds, with its bright red color, crest, and large size.

Illinois, Indiana, Kentucky, North Carolina, Ohio, Virginia, West Virginia

NORTHERN CARDINAL

Cardinalis cardinalis

AMERICAN GOLDFINCH

Spinus tristis

The American goldfinch is also called the wild canary because of its brilliant yellow color.

Whenever they raise two broods in a year, the female builds the second nest while the male feeds and raises the first chicks.

The famous explorers Lewis and Clark were the first to write about the western meadowlark.

This bird builds its nest in a depression in the ground that the female makes with her bill.

Kansas, Montana, Nebraska, North Dakota, Oregon, Wyoming

WESTERN MEADOWLARK

Sturnella neglecta

BROWN PELICAN

Pelecanus occidentalis

Pelicans glide just above the crest of the water, making it look as if they are surfing the waves.

Brown pelicans dive headfirst from as high as 60 feet (18.3 m) to catch fish to eat. When they hit the water, the pelicans scoop fish up in their pouched bills.

Chickadees gather food and hide it for later. When they are ready to eat, they remember where their food is hidden. Black-capped chickadees can remember hundreds of hiding places. These birds eat berries and other fruit, insects, and spiders.

The chickadee is named after its call. It sings *chick-a-dee-dee-dee*, adding more *dees* to the end if there is danger.

Maine, Massachusetts

BLACK-CAPPED CHICKADEE

Poecile atricapillus

BALTIMORE ORIOLE

Icterus galbula

The Baltimore oriole has a bright orange body and a rich, loud whistling song.

The oriole stabs its closed bill into a soft piece of fruit then drinks the juice with its brushy-tipped tongue, a feeding technique called gaping. These birds also eat a lot of insects.

Young loons leave the nest in just one to two days and can dive underwater at two to three days old. Young loons sometimes ride on their mom or dad's back to get around. Parents care for young for about 12 weeks.

Loons are great fliers and swimmers, but it takes them a while to take flight. They run across the top of the water flapping their wings for hundreds of feet in order to get into the air.

Minnesota

COMMON LOON

Gavia immer

Missouri, New York

EASTERN BLUEBIRD

Sialia sialis

Eastern bluebirds nest in man-made boxes, but they don't often show up at bird feeders unless you set out their favorite treat—mealworms.

Hard to miss with their brilliant royal blue back, wings, and head, Eastern bluebirds can often be found on telephone wires and fences.

Its nest is an open cup made of bark, twigs, weeds, and even rootlets. This bird lines its nest with grass, animal hair, and moss.

New Hampshire

PURPLE FINCH

Haemorhous purpureus

The purple finch's big beak and tongue help it crush seeds and remove nuts. These birds also eat insects, including beetles and grasshoppers.

GREATER ROADRUNNER

Geococcyx californianus

Standing two feet (.61 m) from beak to tail, this bird earns its name fairly, as it can outrun a human. It actually prefers running to flying.

When the greater roadrunner runs, it leans forward until its body is almost parallel to the ground. The bird uses its tail as a rudder to steer itself in the direction it wants to go.

Its scissor-shaped tail helps this bird make sharp twists and turns in the air to catch insects for dinner.

A wide assortment of human products are used by the scissor-tailed flycatcher when it builds its nest. These include such things as cloth, string, paper, and even carpet fuzz.

Oklahoma

SCISSOR-TAILED FLYCATCHER

Tyrannus forficatus

Pennsylvania

RUFFED GROUSE

Bonasa umbellus

The ruffed grouse can eat hemlock seeds and other poisonous plants. Hemlock seeds don't harm the grouse, but they do make the bird poisonous to any would-be predators.

At mating time, the male makes a drumming sound by standing on a log and beating his wings in the air.

Like the blue hen, the Rhode Island red is one of only two state birds that is also a domestic bird. It is a breed of chicken.

The Rhode Island red was developed by cross-breeding chickens from Asia with chickens from Italy.

Rhode Island
RHODE ISLAND RED

Gallus gallus domesticus

CAROLINA WREN

Thryothorus ludovicianus

The Carolina wren is known for its distinctive call, which sounds like *teakettle-teakettle!*

These birds use their curved beaks to shake apart bugs to eat. Much of a Carolina wren's diet is made up of spiders and insects.

When flushed from cover, this pheasant flies up almost vertically. It can reach speeds of about 40 miles an hour (64 km/h).

The ring-necked pheasant's rooster-like crowing can be heard nearly a mile (1.6 km) away. It was introduced from Asia into the United States in the 1880s as a game bird.

South Dakota

RING-NECKED PHEASANT

Phasianus colchicus

CALIFORNIA GULL

Larus californicus

Though gulls are often thought of as coastal birds, California gulls spend a good part of their lives inland in places like pastures, scrublands, and garbage dumps.

This bird is known to eat alkali flies, which are found on salty lake shores. The gull catches them by running through a group of flies with its head down and bill open.

The hermit thrush hops about and scrapes leaf litter with its feet when looking for food. Scientists call this behavior foot quivering.

The wood thrush (*Hylocichla mustelina*), a relative of the hermit thrush, is the official bird of Washington, D.C.

Vermont

HERMIT THRUSH

Catharus guttatus

STELLER'S JAY

LISTENING to BIRDS

Birdsong is the soundtrack of our world.

BIRDSONG

If you step outside and listen to singing birds, it may just sound like a symphony of noise. But each species of bird, and even birds within a species, have their own special songs.

A duck's quack doesn't sound like a yellow-breasted chat's *chat-chat-chat*. Barred and great horned owls may both be owls, but they sound completely different. The barred owl's call sounds like this: *Who cooks for you, who cooks for you all*. The great horned owl's call sounds like this: *Whoo-whoo-who-who-who-whooo*. They are speaking different languages. The duck and the chat and the two owls are different bird species. They don't understand each other any better than we humans understand a horse or a lion, both of which are mammals like us. Some birds sound much more alike—can you tell a phoebe from a chickadee? Both say *fee-bee fee-bee*, but the chickadee and the phoebe can certainly tell they are not the same.

In many songbird species, there are differences in their songs based on where they live. Think of this like an accent—people from Boston don't sound like people from Arkansas, but they can still communicate. An example of this in the bird world are the ring ouzels. Ring ouzels in Scotland sing differently than the same birds in Norway. They also sing differently in neighboring valleys. All ouzels have a group of songs (or phrases) in common, but each group living in a region shares unique songs just within that group.

How did scientists figure all this out? They studied baby songbirds raised away from other birds. Without hearing birdsong, the baby songbirds never learned to sing like birds raised around other birds. Scientists also made recordings of wild birds. Carrying large microphones hour after hour, in all kinds of weather, they got up early and stayed up late, looking for birds and recording their songs, which they brought back to the lab to study.

So how do birds learn their songs? Some, like flycatchers, seem to have all the instructions they need encoded in their genes. Other birds, especially songbirds, learn all their songs by listening to adult birds.

COMMON YELLOWTHROAT

LISTENING TO WRENS

One scientist who studies birdsong is Dr. Donald Kroodsma.

Dr. Kroodsma grew up in rural Michigan walking the backwoods, hunting pheasants and rabbits, and listening to nature. He will tell you that he doesn't have better ears than anyone else, but he is an expert at *seeing* birdsong. He does this by recording birds as they sing in the wild and then converting their songs into sonograms. These visual graphs of sound look a bit like sheet music that was smudged before the ink was dry. They hold lots of important information. By both hearing and seeing sound, Dr. Kroodsma can study and learn from the songs and vocabulary of birdsong.

Collecting and analyzing birdsong takes a lot of patience. Dr. Kroodsma is out all hours of the day, especially the hour before sunrise, during what is called the "dawn chorus." He records using various kinds of microphones, a digital recorder, and a computer. Then he studies the sonograms made from his recordings.

How did Dr. Kroodsma get started studying birdsong? After college, he was searching for a project for graduate school. He knew he wanted to study birdsong, but he didn't know what question he wanted to research and, hopefully, answer. Dr. Kroodsma traveled up and down mountains and around the country watching and listening to birds. He found nothing he wanted to make his life's work. Until he returned home.

In his backyard, he heard a Bewick's wren singing. Later, he heard other Bewick's wrens singing elsewhere and they were singing different songs. How did they learn these different songs and from whom? This was the question he wanted to answer.

Dr. Kroodsma already knew many things about wrens. Only the males sing. The fathers sing to their chicks just before feeding them. Wrens, like other birds, have different dialects—they may be similar, but birds of the same species from different areas often sing different songs. He needed to know more.

BEWICK'S WREN

Dr. Kroodsma started to research. Carefully, so as not to hurt them, he caught birds in nets and banded their legs. Later, when he spotted a bird, he could identify it by those bands. For example, on March 20, 1971, Dr. Kroodsma found a Bewick's wren with a band on its left leg and two bands, green over red, on its right. By referring to his notebook, he knew it was bird #326,

How Do Birds Hear?

Birds' songs and calls are important to nearly all species of birds, so it should be no surprise that birds come equipped with excellent hearing. Their ears are similar to those of humans and other mammals, although, unlike humans, birds lack an outer ear so it's harder to tell where their ears are located. Birds' ears are actually slightly below and behind their eyes, below a protective covering of soft feathers. Birds cannot only hear and remember a wider range of pitches, but they hear up to 10 times faster than humans. So while you hear one short note, a bird might hear 10 separate notes in the same sound.

OWL EARS

Most owls have hearing that far exceeds the hearing of other birds. The concave shape of owls' faces acts as an outer ear, funneling the sound into the inner ears that are located at slightly different heights on either side. This creates a tiny difference in the time it takes each ear to hear a noise and helps the owl locate its prey in darkness.

BABY BIRDS

Hearing plays a big role with many birds as soon as—and even before—they hatch. Mother mallards are known to have specific calls they use while incubating their eggs. When a hatchling emerges, it can immediately recognize its mother because of this call. Birds that nest in colonies have the ability to pick their parents' calls out of thousands of others.

EURASIAN EAGLE-OWL

A MOTHER MALLARD WITH HER DUCKLINGS

That Bird Song

That bird song,
the one stitched together
with phrases as sharp,
as honed, as swift
as a pulse of air.

One moment there,
Then gone into memory.
Until the next bird sings
that song.

—Jane Yolen

BLUE JAYS

92

LOOKING at BIRDS

Open your eyes—you never know what might fly by ...

BIRDS ARE ALL AROUND

To see birds, you only need to open your eyes and look. They are all around you. In the woods, on the beach, on a farm, in the mountains, in the desert, and yes, even in the city.

But to really see birds and identify them, you need some equipment. Start with bird identification tools—bird guide books or digital birding apps. Then grab a pair of binoculars, and load up on the three P's: Patience, Persistence, and Practicality.

Patience: Finding birds can take a long while. Yes, you might see a blur flying by. Or something hopping about in the undergrowth. But getting close enough or getting your binoculars on the bird may take a lot of time. You will miss many chances just being too slow—too slow to spot the bird, too slow to raise the glasses, too slow to focus. But keep going.

Persistence: Learning to distinguish one bird from another takes study and memory. Why is one black bird a crow, yet another equally black a raven, and a third a grackle? Why is one flying hawk a red-tailed, another a goshawk, a third a sharp-shinned?

Study the descriptions of wings, beaks, and legs, the color of various parts of the bird's body, and the way the wings are set when they fly. Then learn to put those bird parts together with the bird's name. With this knowledge, you will become an excellent birder.

Practicality: This means when out birding you should have with you your binoculars, bird books or apps, and good walking shoes. Wearing layers of clothing is also important so you can add or shed a layer according to how warm or cold or windy it is outside. Have a backpack with some fresh water and snacks because you may be outside for some time. You also need sunblock to protect from sunburn. Mosquito repellent is another good item to take along, to keep those bugs away. A compass is helpful if you plan to go far from the road, and a notebook and pen with which to write down the birds you see and their descriptions. You may see a rare bird and want to report that to your local birding club or through the eBird app, which will verify the sighting and send out alerts to other birders. And be sure to be with a trusted adult at all times and carry a cell phone in case of emergencies.

Once you have these tools, enjoy the outdoors, look and listen carefully, and have fun!

Bird Tech

The world of technology is changing very quickly. That includes the digital tools you can use to help you become an expert bird-watcher. New ways to find, identify, and record your sightings are being invented every day. Here are some tools that might be helpful. Before using any of these, make sure to ask a trusted adult for permission to download and use these programs.

eBird: On *ebird.org*, you can keep track of your life list, share sightings with other birders, view maps, sign up for rare bird notifications, and contribute, through your sightings, to global conservation efforts.

Birding apps: There are lots of birding apps that you can use to help you identify a bird you see. They range from beginner level all the way to birding expert. Some free ones to start with include these three:
- Merlin Bird ID
- Birdsnap
- Audubon Bird Guide

BROWN PELICANS

HOW TO PHOTOGRAPH BIRDS:

Birds have been a popular subject for artists as far back as the Stone Age, and they continue to fascinate artists of all mediums, especially photographers.

Capturing birds with a camera can be extremely rewarding, but it comes with a certain set of challenges.

The biggest problem with bird photography is that birds are small creatures and often hard to get close to. To fill up the frame and get eye-popping details, you need to close the gap between photographer and bird.

Wildlife photographers have long, high-quality lenses that zoom in on the birds they want to photograph. This allows them to stay far enough away to avoid altering the behavior of skittish species while getting close enough to enjoy every fascinating detail. But photographers have a few other tricks of the trade that you can use, too.

• Choose a location where the birds are accustomed to humans in close proximity: backyards, wildlife refuges, and state or national parks, for example.

• Camouflage or hide yourself. Birds recognize a basic human silhouette and how it moves, so hide yourself against a tree, sit down in tall grass, stay in your car and shoot out the window, or just be still and quiet for a while and become part of the surroundings.

• Let the birds come to you. Pick a spot a bird will come back to time and again and be patient. Locate yourself near a bird feeder, nesting box, or favorite perch, and get comfortable.

Digiscoping

Although photographing birds professionally takes expensive camera equipment and an expert eye, you can get started without either of those. Digiscoping is a way to capture close-up pictures of birds with a digital camera, even a cell phone camera, and a telescope or pair of binoculars.

In order to digiscope, you have to line up the lens of your camera and telescope or binoculars and hold it tight and still. You'll get a full lens shot only if you are lined up correctly. Light leaking in will ruin the shot as will a fidgety hand, so try to stay very still.

It takes some practice to figure it out. But keep at it and you'll be able to get photographs worthy of framing or, at the very least, to help with identification even after the bird you are watching has flown away.

A BIRD-WATCHER TAKING PHOTOGRAPHS OF MIGRATING STARLINGS

How Do Birds See?

Much like humans, birds' sight is their most dominant and important sense, but there are several areas where they differ. First of all, birds' eyes are much bigger than ours when compared to the size of their heads. The increased size and specialized structure create a much larger and brighter image for the brain to analyze. Most birds have binocular vision like us, meaning their eyes are on the front of their heads and the field of vision of each eye overlaps the other. This gives them good depth perception, allowing predators to determine the distance of moving prey while they themselves are moving. Birds that are more likely to be prey often have eyes rotated back on the sides of their head so that they can better sense danger in all directions.

EAGLE EYE

Eagles and other birds of prey have a highly evolved sense of sight that allows them to see objects up to five times farther than the average human. This gives them the ability to spot something very small from a distant perch or while soaring high in the sky.

BALD EAGLE

ULTRAVIOLET COLOR

Scientists have long known that birds can see a broader range and more details of color than humans, but recent studies have shown that most birds also see ultraviolet light (a type of light produced by the sun). This means they can see a whole range of colors that are completely invisible to us.

On the Hunt

You wait as quiet
as leaf or lawn,
or the moment
before dawn.

Endless time,
filling the space.
This breath of air—
your sole workplace.

With a flash,
the hunter's eye
adjusts to catch
the change in sky.

You aim, and click
in perfect light
a wing beat caught,
in frozen flight.

—Jane Yolen and
Heidi E. Y. Stemple

BIRDS on the MOVE

Some birds spend summer and winter in different places. Getting there can be quite a journey.

RED-WINGED BLACKBIRDS

MIGRATION

Many birds move in great numbers across the sky, often in spring or fall. Geese in their vees or hawks circling the air as if the sky is aboil with them, swallows heading ... somewhere. This movement is called migration.

But what exactly does that mean?

In short, bird migration is a regular seasonal movement, often north and south along a flyway, or flight path.

Bird migration has been recorded for a long time. The ancient Greeks Homer and Aristotle mention it. It's even in the Bible.

But do all birds migrate? No. Only about 40 percent of the world's bird species do. That's about 4,000 species.

Some birds migrate long distances between breeding places and their wintering grounds. In fact, half of North American breeding birds leave to find a larger food supply and warm places to spend winter. Some birds migrate only short distances, often if the water in their territory is frozen over during the colder months.

Other triggers for migration are the length of the day (some birds prefer longer days) or genetics—if the bird is born to migratory parents, it will also be a migrant.

Birds that don't migrate are considered permanent residents. They have the ability to withstand the weather where they live, even very cold temperatures, and they have plenty of food sources available all year long.

How Do Birds Know Where They're Going?

Rarely do the routes of migrating birds vary too much. They have a kind of map in their heads. Scientists have found that some birds, like pigeons, actually have a small zone made of a magnetic mineral called magnetite in their brains. This material acts just like a compass, guiding birds in the right direction.

Scientists think that birds map the Earth's magnetic fields through a process called quantum entanglement. The researchers believe molecules found in the birds' eyes play a part in helping birds find Earth's magnetic field, allowing them to sense which way to migrate.

A FLOCK OF MIGRATING SNOW GEESE

LONG JOURNEYS

Thousands of birds migrate all over the world. Here are a few amazing fliers that travel especially long distances.

Arctic tern
(*Sterna paradisaea*)
44,000–50,000 miles (70,800–80,500 km)

Sooty shearwater
(*Ardenna grisea*)
40,000 miles (64,400 km)

Short-tailed shearwater
(*Ardenna tenuirostris*)
27,000 miles (43,500 km)

Northern wheatear
(*Oenanthe oenanthe*)
18,000 miles (29,000 km)

Pectoral sandpiper
(*Calidris melanotos*)
18,000 miles (29,000 km)

Purple martin
(*Progne subis*)
8,000 miles (12,900 km)

Canada goose
(*Branta canandensis*)
2,000–3,000 miles (3,200–4,800 km)

Rufous hummingbird
(*Selasphorus rufus*)
3,000 miles (4,800 km)

Ruby-throated hummingbird
(*Archilochus colubris*)
1,000–1,300 miles (1,600–2,100 km)

THE MISSING SWALLOWS OF
SAN JUAN CAPISTRANO

On St. Joseph's Day, March 19, there is a great festival in San Juan Capistrano in Southern California. Streets fill with a parade, costumed children, dancers, and food, and the historic mission bells ring. The celebration once heralded the arrival of migrating cliff swallows as they returned from their winter habitat in Argentina and nested in the walls of the mission. Legend has it that they started nesting there after an angry innkeeper destroyed their nests. But recently, instead of nesting in the walls of the mission, they have flown past it to raise their chicks elsewhere.

Migrating birds are creatures of habit. Scout birds, the first to arrive before the flock, had been showing up in the area of San Juan Capistrano, likely for centuries, stopping at the largest structure, the mission. It is the perfect place for swallows to nest—especially after an earthquake in 1812 left its eaves exposed. A few days later, the rest of the flock would arrive to rebuild and repair their existing mud nests, in which they would lay their eggs and raise chicks. But, when the nests were removed in the 1990s during a restoration of the mission, the swallows moved on, much to the dismay of their many adoring fans. Bird lovers wanted to bring them back. But how?

Enter Charles Brown, an expert on cliff swallows from the University of Tulsa in Oklahoma. In 2012, Brown set up a recording of swallows at the mission that was played on continuous loop to try to lure the flock back to their nesting spot. This got the birds interested and some flew in to check out the sounds, but none stayed. Since cliff swallows prefer to nest in established nests, Dr. Brown's next plan was to attach man-made nests to a wall, which he did in 2016. Now it's a waiting game to see if this brings the birds back. It could take a couple of migration seasons to find out if it works. But for now, the celebration, filled with hope for the swallows' return, goes on.

MISSION SAN JUAN CAPISTRANO

CLIFF SWALLOW

Vee

The seasons
    are punctuated
    by the comings
    and goings
    of geese.

      Ooh-awnk
      Ooh-awnk

          Fliers-in-line
              draft
                as the front man
              slices
          the headwind.

      Ooh-ank
      Ooh-ank

The Vee flies on,
letting those of us
below,
know.
The season is over.

Another one is on its way.

          —Heidi E. Y. Stemple

BIRDS IN FLOCKS

A flock is a gathering of several birds of the same species, or occasionally birds related but not of the exact same species, to forage for food or to fly.

Why do birds form groups? They understand that they are safer as a group than by themselves.

Through the centuries, many birds received multiple names for their flocks. Some of those names reflect the birds' habits, or are metaphors, or are interesting plays on words. The old English term for such naming is "venery."

The following is a list of some of the most interesting, beautiful, amusing, or spot-on names for flocks. Inside that list are scattered couplets—two line poems—that further play on the flock names.

A FLOCK OF LESSER FLAMINGOS

A **ballet of swans,** no pliés,
In a feathery tutu haze.

Band of jays
Bevy of quail
Brood of hens

A **bouquet of pheasants** flushed from the grass
A blossom of feathers as they pass.

Cast of falcons
Concentration of kingfishers
Cover of coots
Convocation of eagles

A **charm of hummingbirds,** magic spells,
At each wing beat, enchantment swells.

Deceit of lapwings
Descent of woodpeckers

A **dole of doves** sigh,
As winter is nigh.

An exaltation of larks
Fall of woodcocks
Fling of dunlin

Pink feather costumes enhancing
A **flamboyance of flamingos** dancing,

Gaggle of geese

A **kettle of hawks** boils in the air.
Tea party starting—small birds beware.

Murmuration of starlings

A **muster of turkeys** on the move.
The sergeant major will approve.

A **murder of crows,**
In the hedgerows.

Parliament of rooks
Peep of chickens
Pride of penguins

A **paddling of ducks,** a puddling, too,
If you're too close, they'll splash on you.

Quarrel of sparrows.
Sedge of bitterns
Siege of herons

Spring of teal
Squadron of pelicans

A **tiding of magpies** brings bad news
Or brings good word—if you so choose.

Trip of dotterel
Unkindness of ravens
Wisdom of owls.

A **watch of nightingales** in the night
Their songs the only spark of light.

—All couplets by Jane Yolen

SAVING *our* BIRDS

There are about 10,000 known species of birds in the world. They are important to the health of our planet. We must protect them.

BALI STARLING

Gone Forever

Extinction, the complete dying out of all members of a species, is permanent. Though it is possible for a species to become extinct through natural causes, in recent centuries most bird species have become extinct because of humans. We have built on their breeding grounds, poisoned them with our chemicals, and overhunted them for food and feathers. The only thing we can do is to learn from our mistakes and take steps to save other species that are threatened.

MALE PASSENGER PIGEON

FEMALE PASSENGER PIGEON

PASSENGER PIGEON

Ectopistes migratorius

Rock pigeons, also known as rock doves, are found in just about every city and countryside. It may seem that they are safe from extinction, but their close relative, the passenger pigeon, is completely gone. At one point the flocks of passenger pigeons were so thick, the skies darkened with them. In the 17th and 18th centuries, billions of passenger pigeons lived in the eastern United States and southern Canada. Some naturalists estimated that the flying flocks were hundreds of miles long and several miles wide.

In 1813, the great bird artist John James Audubon, while on a horseback trip in Kentucky, found himself beneath a migrating flock. "The light of noonday was obscured as by an eclipse," he remarked in his journal. "The dung fell in spots, not unlike melting flakes of snow."

The hungry birds had healthy appetites. They ate flower and bush seeds, as well as acorns, chestnuts, and beechnuts. And they left a lot of damage behind. The flocks actually uprooted trees because so many of them roosted on the branches at the same time. The birds also smothered more fragile plants in the undergrowth under a layer of droppings several inches thick. So how could those great flocks of passenger pigeons—once the most numerous species of bird in North America—completely die out?

First of all, they were an easily caught source of protein. Hungry colonists as far back as the 1600s killed and ate them. According to the Massachusetts Bay Colony governor John Winthrop, the birds "proved a great blessing." "Multitudes of them," he wrote, "were killed daily." Later, settlers of the American West began building up towns and farms, destroying much of the birds' natural habitat. Loggers cut down forests where the pigeons roosted. There were even professional pigeoners who netted and shot countless numbers of the birds, mostly to sell for food, but sometimes just because the birds were considered nuisances.

The last wild passenger pigeon was shot in 1900. After that, the only living ones were in zoos. Eventually, there was only one left—Martha, named after the wife of George Washington, the first American president. She became a zoo celebrity. People came to take her photograph and paint her likeness.

Martha died of old age on September 1, 1914. You can still see her today on display at the National Museum of Natural History in Washington, D.C.

ROCK PIGEON

Gone Forever

THE DODO

Raphus cucullatus

Isolation almost never makes you safer.

For millions of years the dodo lived on the tiny island of Mauritius in the Indian Ocean, more than a thousand miles off the coast of Africa. It was as isolated as a species can be and was adapted perfectly to its environment.

There were no land mammals on the island, so the dodo didn't have to fly to escape them. Over the millennia, it grew larger and heavier and eventually lost the ability to fly at all. But it wasn't fat and stupid—it was lean and fast, and had a big brain and a keen sense of smell that helped it find food on the ground. But none of its adaptations could save it when the first land mammal finally arrived on the island: humans.

Sailors discovered the island and began stopping there to stock up on freshwater and food. Fresh meat was scarce at sea, so the sailors began eating the dodos. The birds weren't afraid of the sailors and were easily caught and killed. But it wasn't until the Dutch settled the island that the dodos were truly in trouble. The sailors came and went, hunting the dodo, but then leaving the birds alone. When the Dutch settlers decided to stay, they hunted and ate the dodos full-time. And they brought with them more mammals, including cattle, pigs, goat, deer, and accidentally, rats.

All ships have rats. And just like the sailors who'd been cooped up for months at sea, the rats were eager to get back onto land. And what a land it was! Mauritius held no natural enemies for the rats and they had plentiful food in the form of dodos' eggs. With an easy food source, the rats' population exploded. The dodos had no defense against these new predators, and in 1681, just a few hundred years after the first sailor set foot on the island, the last dodo was killed. These birds were perfectly adapted to their isolated island home. When it changed, they were unable to survive.

But what happened to the Dutch? They abandoned the island in 1710 after two failed colonization attempts. The main reason for the two colonies' demise? An infestation of rats. A fitting ending to the story of an accidental introduction of a species.

Gone Forever

HEATH HEN

Tympanuchus cupido cupido

A subspecies of the greater prairie-chicken,
these grouse were common along the east coast of what is now the United States until the colonial era, when they became easy meals for the settlers. By the late 1800s, heath hens were gone except on the Massachusetts island of Martha's Vineyard.

The worried locals outlawed hen hunting and preserved the hen's remaining habitat as best they could. The heath hen population exploded. But in 1916, a wildfire decimated the birds. By 1929, only one male could be found. Booming Ben, named for the sound males made during their courtship dance, returned for three more seasons looking for a mate. But in the end, he disappeared, too. The islanders may not have kept the birds from extinction, but they were one of the first groups to launch an organized attempt to save an endangered species.

Gone Forever

IVORY-BILLED WOODPECKER

Campephilus principalis

Due mostly to habitat destruction,
this striking bird is thought to be extinct. But is it? Ornithologists have been trying to find one in the swamps of the southern United States, where there have been reports of sightings. Some expert researchers and birders claim to have seen them while paddling down rivers along the hard-to-access tracts of land. There are grainy videos and handfuls of other evidence—tree scratchings, holes, and audio recordings. But even after hundreds of trips and thousands of hours searching, there is still nothing conclusive. More researchers are chasing down leads in Cuba. If they finally find proof that an ivory-billed woodpecker lives, conservation efforts can be launched to help save the species.

Introduced Species

It is unfair to say that nonindigenous species (not naturally living in an ecosystem) are the cause of endangering or decimating bird populations. If you really look at the root of the problem, it's almost always humans who are the main problem. The dodos are not the only bird affected by careless humans introducing species that don't belong. Some other harmful introduced (by accident or by design) animals include:

• **Cats:** Often brought in to control a mouse or rat infestation, cats breed quickly and adapt to most environments. The five cats that were brought to Marion Island off the coast of South Africa in 1949 increased to 3,400 by 1977, and as skilled predators, they became a very serious problem for the island's bird population.

• **Pigs:** Often introduced into an environment as a food source or to clear brush, pigs decimate the land. In Kauai, Hawaii, introduced pigs threaten the birdlife by not only destroying the understory of the forests but also competing successfully for food. They also dig wallows that fill with water, creating breeding grounds for mosquitoes, and birds have no defense against the diseases spread by the insects.

• **Snakes:** Exotic pets sometimes escape or, more commonly, are released into the wild by their owners once they are no longer cute or fun. In the Florida Everglades, the Burmese python has taken over. The bird population is being reduced by the snakes, though not as quickly as the mammal population. According to some studies, 99 percent of raccoons, opossums, and rabbits have been killed off by pythons.

Endangered

Many species of birds are in danger of becoming extinct and need our help. Creating laws protecting the birds, their habitats, and their food sources is key. But there are things you can do in your everyday life, too. Living a life mindful of the birds around us is a great way to be a friend to birds. Here are some ideas: maintain a bird-friendly environment, only buy products from companies that are committed to protecting birds, and finally, educate others on these practices. The following stories highlight some birds that are in danger of going extinct.

GREAT GREEN MACAW

Ara ambiguus

The large and stunningly colored great green macaw is native to Central and South America.

This bird is being collected for the pet trade, and its habitat is being destroyed by the cutting down of trees to clear land for farming and mining.

These practices are putting the great green macaw in danger, but conservation efforts are under way to try to save it. A 6,000-acre (2,430-ha) reserve has been created in Ecuador to help this bird and other wildlife by protecting the land on which they live. Within that area, man-made nests are being placed in trees to encourage more breeding. That, along with an effort to learn more about the nesting habits of these amazing birds, may very well be enough to keep them from disappearing.

Endangered

CALIFORNIA CONDOR

Gymnogyps californianus

This giant carrion eater's population is down to just under 500 birds, with a little more than half of those actually living in the wild.

They are in such danger because of poaching (being illegally hunted) and habitat destruction, but mostly because of lead poisoning. Lead, which is very toxic to all living creatures, is getting into the condors from lead bullets used by hunters. When an animal has been killed by a lead bullet and is left behind, condors will eat the dead animal, known as carrion. By eating the animal, the condor also ingests the lead from the bullet. The lead slowly kills condors by getting into their organs. To keep wild condors alive, the birds are routinely captured and treated for lead poisoning by medical professionals.

Laws banning lead ammunition, education about the dangers of lead, and programs offering nonlead ammunition are the leading conservation efforts to help save this regal species.

BALI STARLING

Leucopsar rothschildi

Bali is an island in Indonesia.

Though there are many birds on Bali, there is only one endemic, or native, bird—the Bali starling. Also called Rothschild's mynah, Bali mynah, and jalak Bali, this beautiful blue-masked white bird is Bali's official mascot. But that title did not keep it safe. The bird is so prized in the pet trade that, at one point, there were only about 10 of them left in the wild. So many birds were captured and sold as exotic pets, the species almost disappeared. In 1970, it was placed on the endangered species list but it wasn't until the 1990s that people stepped in, in earnest, to try to save the Bali starlings from extinction. They did this through captive-breeding programs. One such program, housed at a local resort, successfully bred almost 100 birds.

It wasn't easy. Thieves continued to steal the birds, and new enclosures had to be built for breeding programs off the island. Education programs were started to teach children and adults about the importance of conserving these birds. The Bali Bird Sanctuary, which opened in 2006 on the nearby Nusa Penida islands, is a protected area where Bali starlings that were bred at the resort have been released into the wild. Today, the population of Bali starlings is slowly rising and spreading across the protected islands.

Endangered

WHOOPING CRANE

Grus americana

In 1941 whooping cranes' habitat was disappearing.

The birds were also being overhunted, in part for their beautiful white feathers. There were only 15 birds left in the wild, and it was unlikely they would escape extinction. Luckily, conservationists didn't give up on this tall, elegant creature.

The one remaining flock of whoopers grew under the conservationists' watch. But additional flocks—separate from the existing flock—were needed to ensure their survival. The birds also needed to remain wild, which meant they needed to learn to migrate. An idea was hatched: Operation Migration. A small plane would fly along the crane's migratory path with the young birds following.

First the birds needed to be comfortable with the sound of the plane's motor, so recordings of the aircraft were played to the eggs, then chicks. The nestlings were raised by handlers dressed in crane-like white suits (to hide their humanness) and alongside the aircraft to help imprint the planes into their brains (like chicks to mother birds). When it was time for the young cranes to set out on their first migration, the aircraft took off and the whoopers followed their mother plane-crane!

Operation Migration has been successfully leading migration flights of whoopers since 1999. For now, though, the program is on hold while more research is done to figure out if the method is effective.

The whooping cranes are far from safe, but Operation Migration—as well as other efforts, such as the protection of wintering grounds—has begun to help the cranes rebuild new flocks that may save this species.

WHOOPING CRANES FOLLOW A SMALL PLANE AS IT GUIDES THE BIRDS ON A MIGRATION ROUTE.

WHOOPING CRANE

CERULEAN WARBLER

Setophaga cerulea

This migratory bird is in trouble because both its breeding and winter habitats are disappearing.

These small blue warblers prefer old-growth forests with tall, wide trees and lots of undergrowth. They breed in North America and winter in South America. Their traditional habitat in South America is tall evergreen trees found in the forests of the Andes mountains. Many of these forests have been cut down, however, and replaced by coffee farms; coffee grown in the sun is cheaper and easier to tend. These farms are destroying the cerulean warbler's wintering habitat.

But in both North and South America, conservation efforts are in place to help save these birds. Farmers in South America have begun growing coffee under the shade of trees, which are being planted and tended by conservation groups, private landowners, and the farmers on the cerulean warbler's wintering grounds. These trees are now providing a new wintering habitat for the birds.

In North America, conservation groups are caring for the birds by protecting their existing habitats and creating new ones where trees had been cleared.

Success Stories

Luckily, there are some success stories, too. Birds do come back from the brink of extinction to thrive. Breeding programs, laws banning pesticides, changing harmful farming practices, and protecting and creating new habitats are a few ways to help all bird species, but especially those at risk. Alongside those big programs, though, it is also important to have skilled, caring rehabilitators to care for individual birds when they are hurt or sick. Here are some encouraging stories of birds' survival.

BALD EAGLE

Haliaeetus leucocephalus

The bald eagle was selected as America's national symbol in 1782. At the time, there were at least 100,000 eagles nesting in the new United States. But by 1963, the United States had grown, and there were only 487 bald eagles left. The symbol of America was in danger of going extinct.

Like many birds, the bald eagle had lost nesting territory. Waterways were diverted, and forests were cut down. Farmers considered eagles a threat to their livestock, so the birds were trapped, poisoned, or shot. The Bald Eagle Protection Act, signed in 1940, made it illegal to kill, possess, or sell them. But a far greater danger came just a few years later: the pesticide dichloro-diphenyl-trichloroethane, or DDT.

DDT was first used to fight insect-borne diseases in the military. It was so effective and seemed so safe, its used became widespread. But a deadly secret lay in wait for raptors, including the bald eagle. When eagles ate fish that had absorbed DDT, their eggs developed very thin shells that broke when the mothers tried to incubate them. With so few new chicks being born, the bald eagle population in the U.S. plummeted once again. Extinction looked like a very real possibility.

Then, in 1962, the book *Silent Spring* by author Rachel Carson was published. It was about the overuse of pesticides. The book opened the eyes of many to the dangers of these chemicals. Carson testified before Congress about her findings. By 1970, concerns for the environment were spreading countrywide, and a new government agency was formed: the Environmental Protection Agency (EPA). One of the EPA's first acts was to ban the use of DDT. When that happened, the bald eagle began to recover.

After forty years on the list of endangered species, on June 28, 2007, the bald eagle was removed from the list. America's national symbol had been saved.

RAPTOR REHABILITATION

Suzanne Shoemaker had always loved wildlife.

She began her career working in wildlife research and volunteering for a wildlife rehabilitation center. Eventually, Suzanne decided to get a license to open her own rehab center specializing in helping injured raptors—birds of prey.

The Owl Moon Raptor Center (named after Suzanne's favorite book) is run on the property of her home in Boyds, Maryland, U.S.A. She always has more than a dozen hawks, falcons, eagles, and owls living there. During fledging season, when young birds develop wing feathers and prepare to leave the nest, there are more. With proper care, many of the birds that come to Owl Moon heal and are released back into the wild. Other birds are not able to heal enough to live on their own, so they are given to organizations that care for the birds and use them to help educate people about raptors. Suzanne also participates in these education programs.

It takes many people—ordinary citizens, volunteers, and trained rehabilitators, to save injured raptors. The team at Owl Moon does the important work of healing sick and injured birds, but Suzanne and her group of volunteers also enjoy getting to know each bird as an individual.

On any given day, Suzanne might get a call about an injured bird, and July 5, 2014, began that way. A fisherman had spotted a barred owl entangled in fishing line hanging from a tree alongside the C&O Canal. This canal runs through Washington, D.C., and Maryland. The fisherman managed to cut the line but

RELEASING THE REHABILITATED OWL

the owl didn't fly away. He reported the injured owl to the National Park Service, and an officer there called Owl Moon Raptor Center.

Before anyone from Owl Moon could get there, though, the owl managed to fly off, dragging the tangled fishing line with him. The would-be rescuers were worried, but without an owl to rescue, there was nothing they could do.

Two days later, another call came. A couple had been walking along the canal when a rattling in the trees caught their attention. They looked up into the canopy and saw a distressed owl struggling. This time Owl Moon arrived in time. A treetop rescue would need someone who specialized in tree climbing.

The rescuer climbed adeptly up 60 feet (18.3 m) to reach the tangled owl, whose foot was held to the tree by the barbs of a fishing lure. He worked quickly and carefully to remove the hook from the tree and brought the scared owl down to safety. Immediately, the team from Owl Moon examined their new patient. His leg was limp and his foot was so swollen he couldn't open or close it. They removed the hook from the owl's foot and brought him back to Owl Moon to treat his wounds. Luckily, the leg injury turned out to be just muscle strain and fatigue.

The Owl Moon staff regularly cleaned the wound and changed the bandages on the owl's injured foot while he was fed a diet of tasty frozen mice. They did physical therapy with him to help him regain his strength. With the loving patience and expertise of the Owl Moon staff, the owl's injuries healed and he recovered full use of his foot and leg.

In preparation for release, the owl was put through flight reconditioning on a creance line, a long, light cord attached to both legs like a leash, which keeps the owl from flying off before fully healed. Just three months after the owl's rescue, the couple who had spotted the wounded owl had the honor of releasing him back into the wild.

INJURED BARRED OWL AT THE OWL MOON RAPTOR CENTER

OSPREY

Success Story

OSPREY

Pandion haliaetus

Ospreys, though birds of prey like eagles and hawks, are in a family all their own. These birds have special adaptations that set them apart, including a reversible toe and nasal valves that close to keep out water.

Humans almost wiped out ospreys entirely beginning in the 1940s when they started using the pesticide DDT for pest control and to eradicate pest-borne diseases such as malaria and typhus. This is the same chemical that harmed bald eagles.

When the DDT ran off into the water, it contaminated the fish that make up most of the osprey's diet. The combination of DDT and habitat destruction caused by a shoreline real estate boom landed ospreys on the endangered species list.

The DDT ban in 1972, an egg relocation program into foster nests, and building nesting sites where trees were no longer available has succeeded in removing the species from the endangered list, helping their numbers recover and flourish.

OSPREY EGG

What You Can Do To Help Raptors

What can you do to keep birds of prey safe? Encourage your parents to never use poison to control pests (especially rodents) because it kills much more than mice and rats. Raptors will happily eat the poisoned prey and it will make them sick, and they may even die. Tell any hunters and fisherman you know to please use copper or other lead alternatives in their ammunition and sinkers. Lead poisons birds of prey when they eat animals that were shot by lead bullets or fish that have swallowed lead sinkers. Properly throw away any plastic netting, fishing line, or barbed wire that you come across. Raptors can get entangled and become trapped and injured. Unable to move, they will starve if not rescued. Refrain from cutting down trees during raptor nesting seasons (February/March through June/July). Many owls nest in tree cavities and also in some hawk nests that are well hidden in the foliage. If you find an injured raptor, before you try to help it, call a local wildlife rehabilitator. Sometimes what appears to be an injured bird may be a fledgling in the care of its parents that doesn't require rescue. And raptors—especially scared, injured raptors—can be dangerous to handle without the help of someone with experience.

WILD TURKEY

Meleagris gallopavo

Overhunting and loss of habitat almost doomed the wild turkey in the early 1900s. When it seemed that the species might become extinct, a group of people, including President Theodore Roosevelt, stepped in to help. Lands where wild turkeys lived were protected by law. Breeding programs were also started. There was even an attempt to raise turkeys in captivity and release them into the wild, but that effort failed. Those turkeys never became wild enough to survive. Then, a program to trap wild turkeys and move them to new areas was successful—so successful that the 30,000 turkeys (which may seem like a lot but is not) soon became seven million. Today, there are plenty of wild turkeys in north and central America as well as some places in Europe, where they were introduced as game birds.

PIED AVOCET

Recurvirostra avosetta

The avocet is the poster bird for the United Kingdom's (UK) conservation group, the Royal Society for the Protection of Birds (RSPB). Though not endangered worldwide, this beautiful, curved-beak wader was endangered in the UK in the 1940s. World War II was raging, and no one was paying much attention to bird populations. But then a seawall in Suffolk, England, was blown up, and to prevent soldiers from invading, an area nearby was flooded, creating a perfect breeding ground for avocets. These birds began returning to their new and improved UK habitat, and the birds were able to thrive. They were brought back through an act of accidental conservation.

Egg Shells — A Lullaby

Hush my baby,
in your nest
under Mama's
feathered breast.

Man-made nursery,
platform high
where Mama keeps
a watchful eye.

Tucked in safely—
shelled cocoon.
The time to hatch
is coming soon.

Emerge young osprey,
spread your wings.
Enjoy the fish
your Mama brings.

Hush my baby,
in your nest,
under Mama's
feathered breast.

—Heidi E. Y. Stemple

PEREGRINE FALCON

BIRD RECORDS

From the largest to the smallest, these record-holding birds certainly have something to crow about.

Largest
OSTRICH

Struthio camelus

The largest bird is the ostrich, found from the plains of Africa to parts of the deserts of the Middle East. A large male ostrich can reach a height of 9.2 feet (2.8 m) and weigh over 344 pounds (156 kg).

Smallest

BEE HUMMINGBIRD

Mellisuga helenae

As its name implies, the bee hummingbird is about the size of a bee. It can perch comfortably on the eraser top of a pencil. The bee hummingbird is found on mainland Cuba and the Isle of Youth, an island just off the coast of Cuba. From the tip of its bill to the end of its tail, the male is 2.24 inches (57 mm) in length. A male weighs approximately .06 ounce (1.6 g). The female is only slightly larger.

Strongest

HARPY EAGLE

Harpia harpyja

The harpy eagle may take the prize as the strongest bird in the world. In 1990 in Manú National Park, Peru, one carried off a 15-pound (6.8-kg) howler monkey. This was quite a feat as the average harpy eagle weighs only 20 pounds (9.1 kg). Once a well-known resident of the tropical lowland forests of Central and South America, it's now almost gone from the region of Central America because of the loss of habitat.

Fastest on Land

OSTRICH

Struthio camelus

Once again, the ostrich wins. Even though it looks ungainly, it can run at speeds up to 45 miles an hour (72 km/h).

Longest in Air

SOOTY TERN

Onychoprion fuscatus

The sooty tern, a seabird of the tropical oceans, is still a young bird when it leaves the island where it was born. Once airborne, it can remain in flight for 3 to 10 years, taking one- to two-second naps as it goes. Once the sooty tern becomes an adult, it returns to a tropical island to touch down and breed.

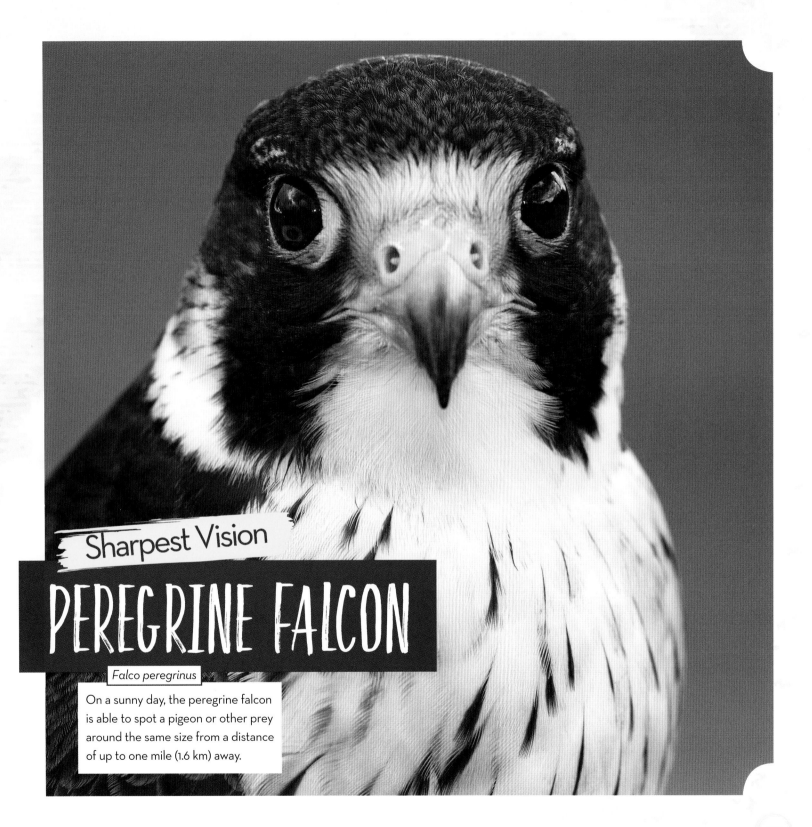

Sharpest Vision

PEREGRINE FALCON

Falco peregrinus

On a sunny day, the peregrine falcon is able to spot a pigeon or other prey around the same size from a distance of up to one mile (1.6 km) away.

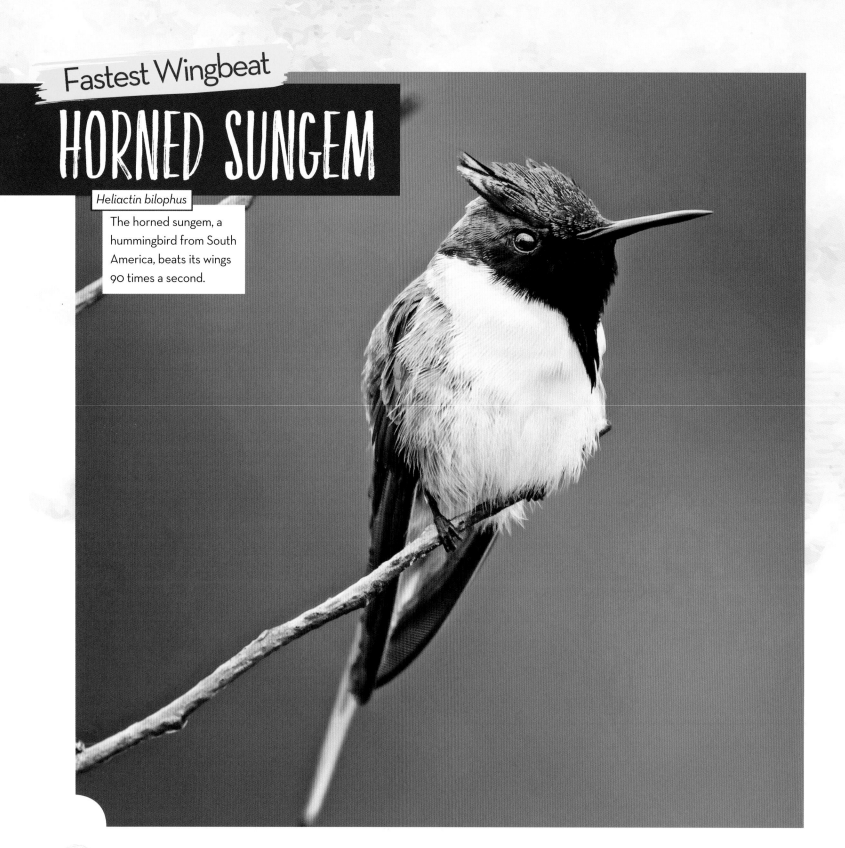

Fastest Wingbeat

HORNED SUNGEM

Heliactin bilophus

The horned sungem, a hummingbird from South America, beats its wings 90 times a second.

AMERICAN WOODCOCK

Scolopax minor

The American woodcock and its cousin the Eurasian woodcock have both been timed flying as slowly as five miles an hour (8 km/h) during courtship displays. At other times, they fly faster.

133

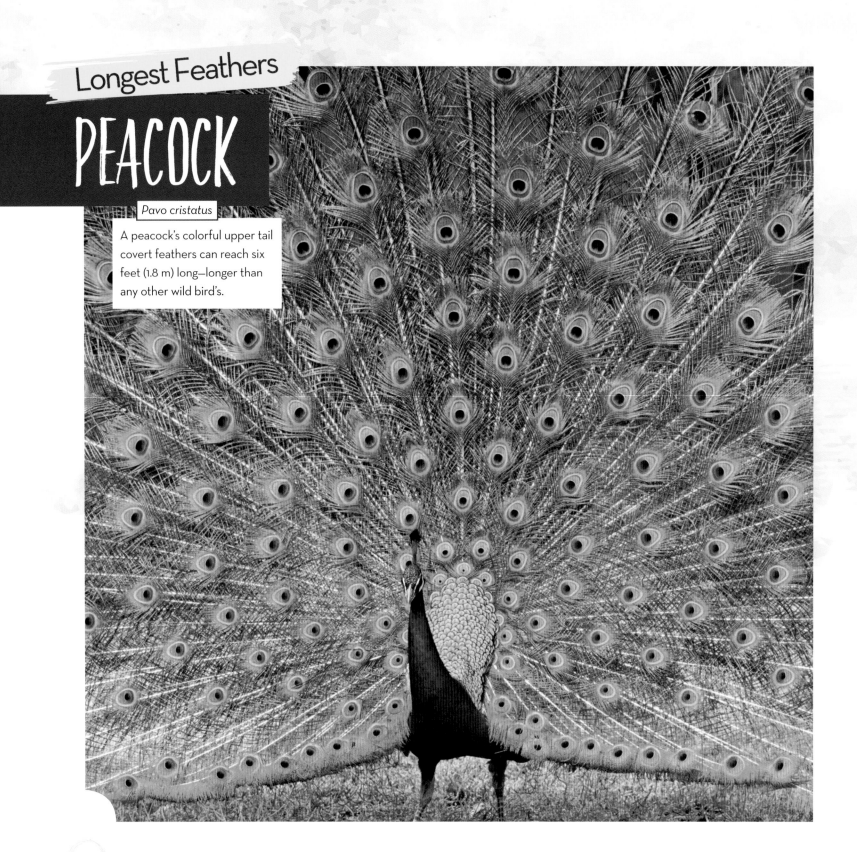

PEACOCK

Pavo cristatus

A peacock's colorful upper tail covert feathers can reach six feet (1.8 m) long—longer than any other wild bird's.

Longest Beak

AUSTRALIAN PELICAN

Pelecanus conspicillatus

At 13 to 18.5 inches (33 to 47 cm), the Australian pelican's beak actually measures the longest. However, the longest beak in relation to body length is that of the sword- billed hummingbird *(Ensifera ensifera)* of South America. At four inches, (10.2 cm) the beak is longer than the bird's body, excluding the tail.

SWORD-BILLED HUMMINGBIRD

135

MALLEEFOWL

Leipoa ocellata

Hands down, the biggest nests are the incubation mounds built by the malleefowl of Australia. These nests can be up to 3.3 feet (1 m) tall and 10 to 16 feet (3 to 5 m) wide. Chicks are on their own from the time they hatch. They dig themselves out of the mounds and can run almost immediately. The young chicks can fly at one day old.

VERVAIN HUMMINGBIRD

Mellisuga minima

The vervain hummingbird builds a nest that is half the size of a walnut shell. This bird is found in the Caribbean countries of the Dominican Republic, Haiti, and Jamaica, and in the Cayman Islands and Puerto Rico.

PAINTING OF A TRICOLORED HERON FROM
JOHN J. AUDUBON'S BOOK *BIRDS OF AMERICA*

BIRDS in the ARTS

For as long as humans have been creating art, birds have been used as subjects. On the walls of prehistoric caves, the vases of ancient dynasties, and the pages and stages of every culture, you will find birds both real and fantastical.

BIRDS IN PRINT:
AUDUBON'S *BIRDS OF AMERICA*

Though John J. Audubon's legacy in the world of birds covers many bases, it began with observation in nature. Most of his boyhood was spent in France. Early on, Audubon loved the outdoors: hunting, fishing, walking in the woods. At age 18, he immigrated to America.

He lived first at his family's Pennsylvania estate, Mill Grove, where 284 acres of untouched land became his studio and laboratory. But over the years, he narrowed his focus to birds, honing his drawing skills and studying the local birdlife.

He helped conduct the first bird-banding in North America. This consists of netting birds, then carefully placing a light band on one or both legs with information about the species, date, and location. Audubon's efforts proved that eastern phoebes returned to the same nesting sites year after year.

While his combined passion for ornithology and art continued to blossom, Audubon also began a successful career in business so that he might provide for his future family. But in 1819, when hard times hit, Audubon went bankrupt. With the business gone and nothing left to lose, he decided to follow his real passion.

A PORTRAIT OF JOHN J. AUDUBON

He set out to paint all the birds of North America.

His journey began down the Mississippi River, where he documented and painted as many birds as he could find.

Often he was away from his young family for months, sometimes living in the woods hunting and fishing for food. He drew portraits for a few dollars, or taught art to plantation owners' children. All the while, he continued his work on the book *Birds of America*.

Audubon returned home with more than 300 bird paintings. He wanted to have them published in a book along with his stories of life on the American frontier. In 1826 he set off for Britain, where his paintings were incredibly well received. Soon, Audubon had enough fans, subscribers, and money to begin the printing process. Because he had painstakingly painted all of his bird portraits life size, the only way they could fit into a book was with special oversize pages. The original paintings were transferred onto copperplate engravings, and those engravings were printed in black and white. After that, each plate was hand-painted following Audubon's originals in an assembly line in which each colorist applied a specific shade. About 150 different colorists worked on the pages.

It took 13 years to make the book, which contains 435 life-size hand-colored prints. Very few books have ever been published in this size, and none are more famous than *Birds of America*. To this day, it's considered one of the greatest art books ever made.

A HAND-COLORED ENGRAVING OF BELTED KINGFISHERS FROM AUDUBON'S BOOK *BIRDS OF AMERICA*

The Bird in Audubon's Picture

The size of the real bird,
The color never dimmed by age.
It gazes out but never blinking,
Staring at you from the page.

Even if the nest is gone.
Or you have never heard it sing;
Though you are centuries past its past,
Then this painting's just the thing.

The bird lives here upon the paper
though extinction's grabbed its claws.
So gaze upon this handsome creature.
Here it's still what once it was.

—Jane Yolen

BIRDS IN MUSIC

You can find songs about birds in every culture, for every instrument, and from every era.

ROOSTER

I Love my Rooster

I love my rooster
my rooster loves me;
I feed my rooster
on a cottonwood tree.
And my little rooster goes
cock-a-dee-do,
Dee doo-dle doo-dle dee
doo-dle dee do.

THIS IS AN OLD SCOTTISH TUNE WITH AMERICAN LYRICS.

Hush Little Baby

Hush little baby,
don't say a word,
Mama's gonna find you
a mockingbird.
And if that mockingbird
don't sing,
Momma's gonna find you
a pheasant's ring.
And if that pheasant's ring's
not gold,
Mama's gonna find you
a swan to hold.
And if that swan
just flies away,
Mama's gonna find you
a turquoise jay.
And if that jay
won't sing along,
Mama's gonna write you
a brand new song.
And if that song
is long and deep,
Mama's gonna sing
you right to sleep.

—Adapted by Jane Yolen and
Heidi E. Y. Stemple

NORTHERN MOCKINGBIRD

THIS SONG WAS ADAPTED ESPECIALLY FOR THIS
BOOK, BUT IS SUNG TO THE ORIGINAL TUNE. THIS
LULLABY IS AN AVIAN VERSION OF THE ORIGINAL.

TURQUOISE JAY

143

FIREBIRD, THE BALLET

One of the most coveted roles for any ballerina is that of the magical red Firebird.

Based on a Russian folktale, this ballet was first performed in 1910 and it catapulted composer Igor Stravinsky to stardom.

The music swells and the curtain rises to reveal a princess and her nine maidens dancing in a walled garden. They are watched by stone guardsmen. The evil wizard Kostchei the Deathless has imprisoned the maidens here in his golden prison. Outside the walls, the lost Prince Ivan searches for food in the stark land where nothing living can be found.

But what is this? The prince sees a flaming red bird flying by. Prince Ivan, starving from days without food, captures the bird and prepares to have a much needed meal, when the bird calls out to him. She will bestow upon him a magical gift if he sets her free. He agrees. As a token of her promise, the Firebird tucks a feather in Prince Ivan's coat. And, with that, the bird leaps away.

But Ivan follows the Firebird to the gates of Kostchei's estate. There, he spies the beautiful maidens, who let him inside. Ivan also sees the princess and falls deeply and immediately in love with her.

Ivan may be afraid, but he is not daunted by the princess's warnings not to look into the evil wizard's eyes. In the garden there are stone statues that were once men who had dared to look. Why this bravery? The prince has a secret—a feather that promises magic.

In order to free his beloved and her maidens, Ivan must fight the wizard's demons. The battle is fierce. Just as the demons are about to win, Ivan remembers the magic feather. He reaches for the feather, and, wielding it like a weapon, he beats back the demons, then calls out for the Firebird.

The magnificent bird enters. She is holding a menacing, magical sword. The demons cry out to their master. It is time for Ivan to face the wizard.

Kostchei the Deathless appears in a burst of smoke, a crown of spikes on his head and a cape swirling and twirling behind him.

The princess and her maidens cower. Ivan is terrified. *Don't look in his eyes,* Ivan reminds himself as he prepares for battle.

But the Firebird, with her wings of flame, flies to the prince and presents the sword to him. With love in his heart and magic on his side, Prince Ivan plunges the Firebird's sword into his enemy's chest. And Kostchei, despite being Deathless, falls dead on the spot. With Kostchei's spells broken by the Firebird's magic, the stone men turn to flesh and dance with the maidens. Ivan has his princess at last, and the Firebird flies off to dance in other kingdoms. One magical red feather is all she leaves behind.

A DANCER FROM THE AMERICAN BALLET THEATER PERFORMING *FIREBIRD.*

Anna Pavlova's Swan

Each night the bird comes so alive,
Her slender arms just like bird wings.
She opens up her small, fair mouth.
And with a viol's voice she sings.

She flutters, and the stage becomes
A pond. You see it blue as blue.
A dancer in a costume surely,
But she is a swan to you.

—Jane Yolen

ANNA PAVLOVA IN HER SWAN COSTUME FOR THE BALLET *THE DYING SWAN*

Anna Pavlova

Anna Pavlova was a Russian ballerina and one of the most famous ballerinas in the world in the late 1800s and early 1900s. She is best known for her performances in a ballet called *The Dying Swan*, which followed the last moments of a swan's life. *The Dying Swan* was specially created for Anna to perform. In 1922 Anna founded the first ballet company to tour ballet performances around the world.

SWANS

BIRDS in POETRY

Emily Dickinson published only a handful of poems before her death in 1886 at the age of 55.

EMILY DICKINSON

While going through Emily's belongings, her grieving sister Lavinia found more than a thousand poems Emily had written. Over the next dozen years, Lavinia made it her life's purpose to ensure her poet sister would be published, known, and beloved forever more.

Emily wrote about many subjects, but the natural world around her was one of her favorite topics. It is no surprise that birds show up on more than a few occasions in her writings. Here are two of her poems about our feathery friends.

AMERICAN ROBIN

In The Garden.

A bird came down the walk:
He did not know I saw;
He bit an angle-worm in halves
And ate the fellow, raw.

And then he drank a dew
From a convenient grass,
And then hopped sidewise to the wall
To let a beetle pass.

He glanced with rapid eyes
That hurried all abroad, —
They looked like frightened beads, I thought;
He stirred his velvet head

Like one in danger; cautious,
I offered him a crumb,
And he unrolled his feathers
And rowed him softer home

Than oars divide the ocean,
Too silver for a seam,
Or butterflies, off banks of noon,
Leap, plashless, as they swim.

The Robin.

The robin is the one
That interrupts the morn
With hurried, few, express reports
When March is scarcely on.

The robin is the one
That overflows the noon
With her cherubic quantity,
An April but begun.

The robin is the one
That speechless from her nest
Submits that home and certainty
And sanctity are best.

BIRDS ON SCREEN

Birds are notoriously difficult to train to perform on stage or screen.

Most of them are flustered and fluttery around the enormous number of people who work on a stage, or on a movie or television set. Plus, they can be easily confused by the lighting and the constant noise. Also, they will unapologetically let go of bird droppings on the set, lights, props—and on other actors. There are some trainable birds, including crows, ravens, various parrots, and raptors.

Over the years, birds in movies and on television have been portrayed in cartoons, by marionettes, Claymation, soft puppets, and even by actors in bird suits.

Birds have also been portrayed by the use of actual footage of real birds. The footage is manipulated so that the birds seem to be dancing, singing, and interacting.

HERE ARE 9 BIRD MOVIES TO WATCH—IN A BIRDATHON!

EMPEROR PENGUIN CHICKS FROM THE MOVIE *MARCH OF THE PENGUINS*

1. THE EAGLE HUNTRESS

In this prize-winning documentary, a Mongolian teenager, 13-year-old Aisholpan, trains to become the first female in her family to become an eagle huntress.

2. RIO

A bird rescue fantasy, this 3-D computer-animated movie follows a macaw and his friend as they make their way out of captivity.

3. CHICKEN RUN

In this whacky stop-motion animation, an American rooster and a British hen fall in love and plot to escape to freedom, away from a mean farmer.

4. HARRY POTTER

A snowy owl is a partner in magic in this fantastical film. The film used several trained male owls, although the owl in the book is female.

5. MARCH OF THE PENGUINS

This stunning documentary portrays the life of emperor penguins, which live in one of the coldest, windiest places on Earth—Antarctica.

6. LEGEND OF THE GUARDIANS: THE OWLS OF GA'HOOLE

In this computer-animated fantasy adventure, a young owl named Soren helps defend his fellow owls against evil.

7. THE LEGEND OF PALE MALE

This warm-hearted documentary tells the story of a red-tailed hawk living in New York City.

8. VALIANT

A World War II carrier pigeon wants to be a hero in this fun and courageous computer-animated movie.

9. HAPPY FEET

In this Oscar-winning computer-animated musical, a young emperor penguin has a talent for tap-dancing.

PENGUINS FROM THE MOVIE *HAPPY FEET*

BIRDS *in* STORY

RAVEN BRINGS THE LIGHT
Coast of Alaska (United States)
and British Columbia (Canada)

WHY THE BUZZARD WENT SOUTH
Southern
United States

NORTH
AMERICA

PACIFIC
OCEAN

ATLANTIC
OCEAN

SOUTH
AMERICA

Birds appear in hundreds, even thousands, of folk and fairy tales from around the world. These stories still live centuries beyond their original tellings and places of birth. Here are seven multicultural tales from across the globe.

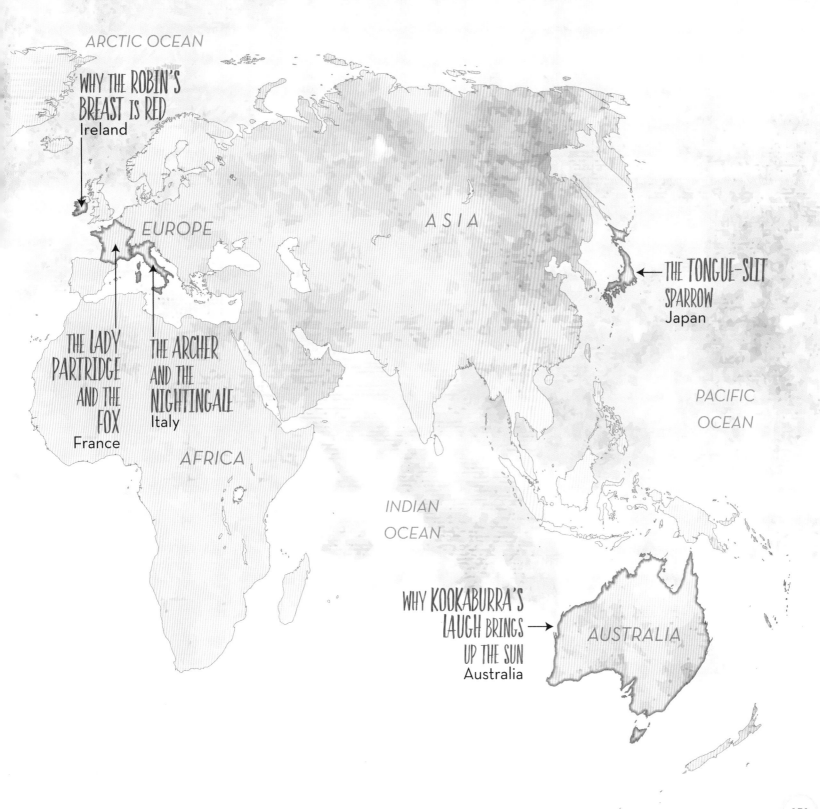

ARCTIC OCEAN

WHY the ROBIN'S
BREAST is RED
Ireland

EUROPE

ASIA

THE TONGUE-SLIT
SPARROW
Japan

THE LADY
PARTRIDGE
AND THE
FOX
France

THE ARCHER
AND THE
NIGHTINGALE
Italy

AFRICA

PACIFIC
OCEAN

INDIAN
OCEAN

WHY KOOKABURRA'S
LAUGH brings
UP THE SUN
Australia

AUSTRALIA

THE ARCHER AND THE NIGHTINGALE

Once a proud archer caught a nightingale in his nets. He was about to wring her neck, when she spoke to him in his own language.

"Why kill me? I am too small to eat, too small to do you harm. If you let me go, I will give you three rules for life. Follow them, you will be a rich fellow indeed."

Astonished at hearing the bird speak, the archer promised to let her go if she told him those rules. He loosened his grip but still kept her caged in his massive hands.

The little bird began: "First, never attempt the impossible; second, do not regret the irrecoverable; third, never believe the incredible. Follow these rules, you will indeed find all the riches you desire."

The archer opened his hands and let the nightingale go.

No sooner was she free and safe on the bough of a tree, that the nightingale called down, "What a silly fellow you are. In my body lies a pearl larger than an ostrich egg. It would have made your fortune."

COMMON NIGHTINGALE

Furious at being fooled, the archer immediately spread his nets to catch her again. But now knowing what they were, she easily eluded them.

So the archer tried promises and flattery. "Come into my house, beautiful singer," he called. "I will feed you with my own hands and let you fly outside whenever you wish."

The nightingale laughed, a sound as lovely as its song. "You are a fool, archer, for you have paid no mind to my three rules. First, you have attempted the impossible—capturing me a second time; second, you have regretted losing the irrecoverable—a pearl of great size; third, you believe the incredible, that my body could contain a pearl three times larger than I am."

She laughed again, a cascade of notes. Then she sang him this song.

Fool you are,
Fool you will remain.
Your chances will not
Come again.

Away she flew.

And indeed, the archer never saw her another time, nor did he ever have another shot at success.

This tale is from the Romans, who ruled for hundreds of years in what is now Italy. Eventually, they controlled an empire that spanned most parts of Europe. The story is from a book called *Gesta Romanorum*, or *Stories From the Romans*.

153

RAVEN BRINGS THE LIGHT

In the beginning the world was totally dark. The only light existed in a tiny box within many boxes that an old man kept in the home he shared with his daughter.

Now Trickster Raven, Shape-Changer Raven—who had existed from the beginning of time—was tired of the dark world because he was forever bumping into things: trees and tree stumps, rocks and crags.

Coming upon the home of the old man in the dark, Raven learned, through his sly ways, about the tiny box and the light, and he made a secret promise to steal it.

So Raven turned himself into a single hemlock needle and dropped into the river.

In the dark morning, the old man's daughter came to the river as usual to dip her basket into the water. Thirsty from a long sleep, the girl drank from the basket and swallowed the needle that was Raven. The needle slid down into her belly.

There Raven changed shape again, becoming a tiny egg-child. And nine months later, he came into the world, this time as a baby. He was an odd-looking boy with a thatch of black hair.

His grandfather loved him. Still, he warned Raven: "Do not touch my treasure box or you will be severely punished."

The boy begged and pleaded, wheedled and whined. And at last his grandfather gave in.

Opening box after box till he came to the smallest one, the old man drew out the glowing sphere of light and threw it to his grandson.

As the light whizzed toward him, Raven turned back into his giant raven form, wings spread wide. He plucked up the glowing light with his beak. Then, pumping his great wings, he flew through the smoke hole in the roof.

And that is how the universe got its light.

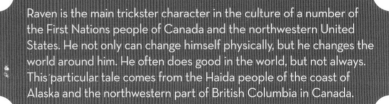

COMMON RAVEN

Raven is the main trickster character in the culture of a number of the First Nations people of Canada and the northwestern United States. He not only can change himself physically, but he changes the world around him. He often does good in the world, but not always. This particular tale comes from the Haida people of the coast of Alaska and the northwestern part of British Columbia in Canada.

WHY KOOKABURRA'S LAUGH BRINGS UP THE SUN

Once in the Dreamtime, when only moon and stars lit the Earth below, and most animals and birds and other creatures had trouble seeing in the dark, the sky spirits decided to make a great fire so the creatures could have both warmth and light. They called this fire "Sun."

But they needed some way to announce this great new gift of theirs. First they tried asking Morning Star. But though Morning Star sparkled and shimmered and tried as hard as it could, no one below seemed to notice. It was too dark.

"We need a big noise," the sky spirits decided. "To announce this new thing, this Light, this Sun." They looked at all the creatures below who might become such an announcer. But they all seemed too quiet, too small.

Just then, Kookaburra—who was sitting on the branch of a gum tree—spotted a mouse. Hard to do in the dark. He flew down and caught it—even harder to do in the dark—and he laughed in delight. It was a huge, raucous sound. No one in all the world at that time was so loud. So brassy. So ... so ... trumpetlike.

The spirits visited Kookaburra that evening in his gum tree. Well, it was actually daytime but still dark. "Be our announcer," they said. "Help us bring up the sun every morning."

Kookaburra shrugged. "Why should I? I'm happy just flying about. Happy eating and flying and eating some more. Doing that in the dark is hard enough. Why should I bother with another job?"

"Then we will take back the gift of Sun," said the spirits, "and leave the world in darkness forever."

Kookaburra thought, *If there is darkness forever, I will not be able to see to fly. I will not spot my food.* Then he chuckled. *But if I say yes to this, I will be a hero. And after all, it will only take a very little time in the morning. The only thing I have to do is laugh.*

And from that day till this, whenever the sun rises in Australia, the kookaburra and his children and grandchildren and great greats, many times over, announce loudly, raucously, and firmly that it is, truly, morning and time for everyone to rise.

LAUGHING KOOKABURRA

This Australian tale is from the indigenous people of Australia. It is a porquoi tale. *Porquoi* means "why" in French. Porquoi tales explain origins, in this case why the laughing kookaburra sings early in the morning. The laughing kookaburra is also called the "bushman's clock."

156

157

WHY THE ROBIN'S BREAST IS RED

Long ago in Ireland, when poverty gripped the land, an old soldier and his young son were traveling along, looking for work.

It was a bitter winter, and they'd wrapped rags around their shoes to keep their feet warm. A cruel wind and a scrim of snow made the night even more dangerous.

The father had thought they might find a cottage, or at least a barn, where they could sleep that night, but they found nothing, not even a hollow tree to shelter in. All they came upon was a large bush, its leaves shed long ago. Still it was the best on offer, so the father built a fire and told his son to try to sleep a little.

The boy managed several hours while his father watched the blaze, for only that kept them warm, kept away the wolves.

But at last, exhausted, the father began to falter. He couldn't keep his eyes open for another minute. So he woke his son, and asked him to watch the fire if only for an hour.

The poor boy tried to keep awake. He all but propped each eyelid open with twigs. But exhausted from the days on the trail, his head drooped.

EUROPEAN ROBIN

As his head got lower, the fire got lower, too. Soon a starving wolf inched nearer to the banking flames, closer and closer to the sleepers. But in the bush, wide awake, was a tiny bird, gray as a snow cloud, gray as the limbs of the briary bush. He hopped right up to the fire.

There he fanned the flickering embers back into flames with his wings. Even when those flames began to touch his small breast, searing his feathers, he did not stop. He kept fanning the flames until morning when father and son awoke, refreshed, and took over the fire.

A bit ashamed, the boy admitted he had fallen asleep. Then he noticed the bird. "Look, Papa," he said, "the bird's breast has turned orange-red."

"It has been burned by the flames," Papa explained.

"It shines like a medal of honor, Papa." The boy knew medals and was just beginning to understand honor, too.

Though they had only a half loaf of bread for the rest of their journey, they left all the crumbs for the little bird. Then, with the good sleep, the half loaf, and the display of selfless courage the little robin had shown them, they made it to Dublin where they both found work.

From that day on, robin has worn his orange-red breast with pride. It is a reminder to everyone how even the smallest can help the strong, as he did—risking his life to save father and son in that deepest of winter.

Like the laughing kookaburra story, this is another kind of porquoi or "why" tale. It can also be read as a parable, or teaching story, where something is done not for a reward but because it's the right thing to do. An important note: the European robin is not a large thrush like the American robin. The European bird is the size of a sparrow and has gray-brown upperparts, a whitish belly, and a face lined with gray. This story is from Ireland.

WHY THE BUZZARD WENT SOUTH

Now Buzzard was visiting his good friend Gull, and they got to talking, as birds do.

"You sure come from a busy family," Buzzard said to Gull. He watched as, overhead, Gull's aunts and uncles and cousins zipped across the sky, before coasting down and grabbing up food from both sidewalk and sea.

"We sure are," said Gull. "We all know how to hustle." Then he added, "You know the old saying, he who hustles makes a living. Only the buzzard waits for something to die."

Buzzard shook his head. "We prefer to say we wait on salvation."

Gull gave a long thought, then said, "I guess both are about the same thing."

Buzzard gave a chuckle. "Though sometimes I think you boys have the right idea." And he settled in like he was about to tell a story. So Gull waited, and sure enough the story came.

"One day," said Buzzard, in that low, confiding storytelling voice, "I smelled the good smell of something that had died. It was a cow that got hit by a train."

"That's a fine smell," said Gull.

"To you, my friend, maybe. Not to me. I like my meat seasoned with time." He meant he liked it old and smelly. "That cow just gave out the new-dead smell. So I went off to wait till it got good and ripe."

Gull chuckled. "What you call ripe, I call stinky!"

Buzzard gave a clatter of his beak. "Stinky was what I was going for, but that cow was gone."

That made Gull stop chuckling. He shivered. "You mean like a zombie cow? I never heard of any such thing before."

"That's because there isn't any such thing as zombies," said Buzzard. "Dead is dead. What I meant was, those railroad section hands had buried that cow so deep in the field, why, even I couldn't get a smell of it."

"That's what happens when you don't hustle," said Gull.

"And that's why I'm heading south," said Buzzard, "where it's often too hot for them to bury their dead till way into the night."

And that's why—to this day—Buzzard makes a good living down in the South.

> This kind of humor tale, sometimes called a thigh-slapper, is meant to poke fun at the tellers and their neighbors. This one is from the African-American tradition.
>
> The buzzard referred to in this story is also called a turkey vulture (*Cathartes aura*).

RING-BILLED GULL

THE LADY PARTRIDGE AND THE FOX

Oh once, *mes enfants*, my children, there was a lady partridge, and wasn't she pretty, with her dusky coat and especially her two red feet.

There she was, perched on a rock, singing her heart out, looking over the heather, singing *"p-tuk tuk tuk cheer ..."*

There came a fox. A crafty fox. A hungry fox. "Why, my lovely Lady P," Fox said, "you have the voice of an angel. The nightingale is a poor singer compared to you." He licked his lips, which she took for deep thought. "But your sainted Mama, so early departed, bless her, had a lovely voice too, only hers was ...."

"Hers was what?" Partridge asked eagerly.

"Hers had a particular moving quality. A purr." Fox's own voice was now almost a purr.

"A purr? But we partridges do not purr, we *tuk-tuk-tuk.*"

Fox shook his head. "It was that purr that was so beautiful. But now that I think of it, the dear lady always closed her eyes when she sang. Perhaps that was what made her voice so ...."

"Moving?" asked Partridge. And wanting a voice as moving as her mother's, Partridge foolishly closed her eyes and opened her mouth to sing.

In one leap, Fox was on her and she was imprisoned in the jail of his mouth, his teeth as strong as iron bars.

But if Fox was clever, he'd misread Lady Partridge, for she was cleverer still. She calmed her trembling limbs and waited for her moment.

As they passed over a bridge, washerwomen in the stream below cried out, "Look, look—there goes crafty Fox with a partridge. There'll be a good supper at his house tonight!"

At this, Partridge said, "If I were you—though clearly I am only a poor bird for your cook pot—I'd shout right back at those silly washerwomen and say, 'A better dinner than you, *pauvre femmes antiques—poor ancient women!'"

Fox loved being cheeky. "A much better din ..." he began. But the moment he opened his mouth to shout at the old women, Partridge flew out of the trap of his teeth and perched on a high branch.

"Tuk—tuk-tuk!" she called, her eyes wide open.

Knowing he'd lost his dinner, Fox was still gracious enough to call up to her. "You've taught me a lesson I shall not forget: Never talk unless it's absolutely necessary."

She laughed and called back, "You have taught me something, too, old Fox—to only close my eyes when I go to sleep."

RED-LEGGED PARTRIDGE

This story comes from France and is related to an Aesop fable from Greece called "The Fox and the Crow." In that one, though, Fox flatters a crow and when he does, she drops a piece of cheese.

MEDIUM GROUND-FINCH

CITIZEN SCIENCE

You don't need a degree in ornithology to become a bird expert.

FAMOUS CITIZEN SCIENTISTS

WHAT IS A CITIZEN SCIENTIST?

A citizen scientist is a regular person who has passion, patience, imagination, and the desire to learn new things about the world around them.

Three of the most famous citizen scientists, ordinary people who changed the way we understand the world around us through their observations, writing, and art, are Charles Darwin, Mary Anning, and Roger Tory Peterson. Their contributions to science and ornithology are immense.

CHARLES DARWIN (ENGLISH, 1809-1882)

As a boy Charles collected beetles and bugs. Though he was supposed to become a doctor like his father, his passion lay in exploring and learning about the natural world. Eventually he became known as a naturalist—someone who spent time in a variety of outdoor studies but never earned a degree in any of them.

As a young man, Charles worked on a small ship, the H.M.S. *Beagle,* voyaging around South America. He was not the ship's naturalist, but companion to the ship's captain. Captain Robert Fitzroy wanted an educated man aboard to have conversations with.

Charles and Robert spent two years sailing around South America, often landing at barely accessible places, or sheltering within rowboat's reach of the islands off the South American coast. Charles collected many species of animals, birds, insects, and sea creatures, making detailed notes and drawings of what he saw.

As the *Beagle* journeyed farther, they stopped for five weeks in the Galápagos Islands, off the coast of Ecuador. This is where Charles found finches. First, he studied these small birds in the wild, then he collected specimens and took them back to England. His observations of these finches became the bedrock on which he built his groundbreaking theory of evolution. One of the finches Darwin studied, the medium ground-finch, is pictured on the previous page.

MARY ANNING (ENGLISH, 1799–1847)

She sells seashells by the seashore. Mary Anning, the subject of that famous tongue twister, was one of the most important citizen scientists in the field of paleontology—the study of fossils. Although Mary never knew it in her lifetime, fossils of dinosaurs would one day help prove that birds are direct descendants of large feathered dinosaurs.

Mary grew up, uneducated, in a small English seaside town. Her carpenter father had a business selling "curiosities," ammonite fossils of coiled mollusks from the Jurassic period, that he found on the local beaches.

Mr. Anning taught Mary and her brother how to find and extricate these fossils from rocks, and how to clean and ready them for market.

When Mary was 10 years old, her father died and she and her brother took over the business. She soon became a well-known fossil hunter.

At age 12, Mary found the first complete skeleton of an ichthyosaur, a large ancient marine reptile that had lived 250 million years ago. When she was 25, she found an almost complete skeleton of another ancient marine reptile, the snake-necked plesiosaur.

Four years later, in 1828, Mary discovered the bones of a *Pterodactylus*, which she called "flying dragon." It was one of the ancestors of modern birds.

Until she died at age 48, Mary found and sold fossils to paleontologists and professors. Although she had almost no education, her knowledge was in her hands and eyes. She could "see" bones hidden in stone before the scientists could. Her findings provided a lot of the evidence for the 19th century's ideas about Earth's history and its fossil layers. And eventually, some of her discoveries helped scientists build a case for the dino-bird connection.

ROGER TORY PETERSON (AMERICAN, 1908–1996)

Roger Tory Peterson went to work in the mills of Jamestown, New York, at age 10, after his Swedish immigrant father died. His German mother was a schoolteacher, but her income alone wasn't enough to support the family, so Roger had to help out.

When not working, Roger enjoyed being outdoors. He loved nature and became fascinated by birds. He joined a Junior Audubon Bird Club at age 11. Soon after, he bought field glasses, a camera, and a bird guide by Chester A. Reed. Published in 1905, Reed's *Bird Guide* was the first guidebook of its kind.

Roger took careful notes of his observations, learning both the common and scientific names of birds. Soon he began to write about birds. In 1925, at 17, Roger's writing was published in *Bird-Lore* magazine. His article, called "Notes From Field and Study," detailed two bird sightings—a Carolina wren and a tufted titmouse.

He was also interested in painting birds, but he needed to learn art techniques. So, in his early 20s, Roger took painting classes at the Art Student's League in New York City.

By the early 1930s, Roger had become so skilled at identifying birds and their small distinctive features (known as field marks), that he wrote his own field guide: *A Field Guide to the Birds*, featuring birds of the eastern United States. It became the go-to guide for both beginning and experienced birders. Later, he wrote a second guide to birds of the western United States. There are still millions of copies of his two bird guides in print.

Even though Roger had no formal education, he was so knowledgeable that he eventually joined the staff of the National Association of Audubon Societies. He also became the art director of *Bird-Lore*, which later became *Audubon* magazine. He revolutionized birding for amateurs with his easy-to-read bird guides and become known as the father of the modern bird guide.

BECOMING A CITIZEN SCIENTIST

There are many ways to become a citizen (bird) scientist. To start, become a bird-watcher. Learn all you can about the birds you see and hear. Study the birds that need help from humans and research ways you can help those species. Then, join a bird club—there's probably one near you.

Bird clubs sponsor events like lectures on birds and birding, owl prowls, hawk watches, shorebird treks, and other field trips and activities. Sign up for bird counts or birdwatches in your area. Fall migration is often a time when many of these groups go out.

The most famous and complete bird counts are:

1. **The Audubon Christmas Bird Count (Western Hemisphere)**
2. **The Great Backyard Bird Count (Worldwide)**
3. **Project Feeder Watch (North America)**
4. **Nest Watch (United States)**
5. **Celebrate Urban Birds (North America)**

You can find their websites listed in the back of this book. There is also a Citizen Science League, which is also listed in the back of this book. There are also a number of online pages filled with information about birds. Begin with National Geographic, Cornell University's Lab of Ornithology, and the main website of the Audubon Society.

HYACINTH MACAW

Do No Harm

This land once
belonged to the birds
who took to the skies,
the oceans,
and back to the land

where they were met by Man—
his nets, his traps, his guns,
his hoes, his scythes,
his land-clearing machines,
his kites, his lines, his planes,
his power.

What was once theirs
was now his.
To own.
To use.
To rule.

To protect.

This land still
belongs to the birds.
Who take to the sky,
the oceans,
and back to the land

where, now,
they are met
by Conservationists,
their laws,
their protections,
their programs.
their care.

What is ours,
is now theirs.
To live.
To flourish.
To sing,
To fly.

—Heidi E. Y. Stemple

171

GREAT TIT

BIRDS in YOUR BACKYARD

There are so many birds to discover;
just take a look in your backyard.

A GULL STANDS ON A TOWER IN ISTANBUL, TURKEY.

BIRD-FRIENDLY HABITATS

Wherever you live, there are birds, and every habitat has its own species. No matter where you are, there are ways to make your backyard or other open areas more bird-friendly.

IN THE CITY

Setting out small feeders with seeds can attract birds to your balcony or window. To attract hummingbirds, you can hang nectar feeders. To ensure none fly into the glass, use feeders that hang right in front of or on the window. This will take some patience—and while you're waiting, make sure to change your seed often so it's ready for the birds when they show up.

If you have a balcony, hanging plants provide nesting spots for birds. Make sure the plants aren't near a large drop-off, to protect nestlings from fledging too close to the edge.

IN THE GARDEN

There are many plants you can grow that attract birds to your garden. Make sure you choose plants that are native to your area. They are best for your garden and best for the birds.

Honeysuckle, bee balm, hummingbird bush, and hollyhock are hummingbird favorites. Honeysuckle also attracts thrushes, waxwings, and purple finches.

Holly berries are great for song thrushes, woodpeckers, thrashers, and catbirds. They eat these berries in the winter.

The fruit of the elderberry bush brings in red-eyed vireos and brown thrashers.

Dogwood trees, known for their flowers, also produce fruit that is eaten by bluebirds, cardinals, mockingbirds, and catbirds.

Strawberry plants are regularly visited by cedar waxwings, among other birds.

Coneflowers are a favorite of bluebirds and goldfinches.

And if you leave sunflowers up to form large seed heads in the fall, the center of tightly packed seeds will entice such birds as goldfinches, cardinals, chickadees, nuthatches, and red-bellied woodpeckers.

MALE EASTERN BLUEBIRD

175

WESTERN SCREECH-OWL

Birds Outside Your Window

Would you like to invite a variety of birds to your backyard? Birdhouses are a great way to do so.

Here are nine birdhouses and the types of birds they attract. If you prefer to make a birdhouse rather than buy one, the most important dimension is the diameter of the entrance hole. If the hole is too small and the bird becomes stuck going in or out, it may damage its feathers. Too big, and a predator bird may use the house, or other predators such as rats, snakes, or raccoons could get in and possibly destroy eggs or kill nestlings. So be sure to make the opening of your birdhouse an appropriate size for the birds that visit you.

PURPLE MARTIN HOUSES:

These birds like either multi-compartment houses or hollowed out gourds strung up next to each other. Purple martin houses need to be 30 to 120 feet (9.1 to 36.6 m) away from human houses and 40 to 60 feet (12.2 to 18.3 m) from trees. They should also be in the open, near water. Choose the center of the largest open spot available. It's best to close the entrance to the house between martin nesting seasons to keep other birds from moving in. The diameter of the entrance hole should be between 1 3/4 and 2 1/4 inches (4.4 and 5.7 cm).

A PURPLE MARTIN COLONY

SCREECH-OWL BOX:

Attach these boxes 12 to 40 feet (3.7 to 12.2 m) high near a wooded area to attract these tiny owls. The entry hole should be 4 inches (10.2 cm) wide by 3 inches (7.6 cm) high. Don't forget the bedding—both eastern and western screech-owls like a couple of inches (5 cm) of wood shavings inside.

BLUEBIRD BOX:

To keep other birds from nesting in your simple wooden covered bluebird box, make sure the hole is 1 1/2 inches (3.8 cm) in diameter for eastern bluebirds and 1 9/16 inches (3.96 cm) for mountain and western bluebirds. The box should be placed 5 to 10 feet (1.5 to 3 m) high on a post facing an open field.

EASTERN BLUEBIRDS

WOOD DUCK BOX:

Place wood duck boxes near wetlands anywhere from 3 to 5 feet (0.9 to 1.5 m) above open water or on a tree 12 to 40 feet (3.7 to 12.2 m) high. The entry hole should be 4 inches (10.2 cm) wide and 3 inches (7.5 cm) tall.

WOOD DUCK

BARN OWL BOX:

A barn owl box needs to be close to good rodent territory—its favorite food. Place the box up 8 to 25 feet (2.4 to 7.6 m) high. It can be outside or in a barn. The entry hole should be 6 inches (15.2 cm) wide by 12 inches (30 cm) tall.

BARN OWL

AMERICAN KESTREL BOX:

You can place these boxes just about anywhere these fierce small raptors live. Just don't put them close together. The boxes need a half mile (0.8 km) between them because the kestrels are very territorial. The entrance hole should be 3 inches (7.6 cm) in diameter.

FEMALE AMERICAN KESTREL

AMERICAN ROBIN

ROBIN AND PHOEBE SHELF:

No fancy house needed for these birds. Robins and phoebes prefer simple nesting shelves. Shelves can be placed on the side of an arbor or tree. They should be in an area surrounded by trees and grass.

NORTHERN FLICKER HOUSE:

Set up this house near meadows, pastures, orchards, or fields in a generally sunny spot. Fill the box completely with wood chips and/or shavings to welcome a northern flicker. The entrance should be 2 1/2 inches (6.35 cm) wide, and the house should be mounted 6 to 20 feet (1.8 to 6.1 m) high on a tree trunk.

NORTHERN FLICKER

WREN BOX:

Both Carolina and house wrens love nest boxes. These birds are tiny and only need a hole about 1 1/8 inch (2.86 cm) to climb in and out of. You should place the boxes about 5 to 10 feet (1.5 to 3 m) high on a post or hanging in a tree.

HOUSE WREN

BIRDBATHS

Water is essential for birds to drink and bathe in.

Having a birdbath in your garden is a great way to attract birds and provide them with the water they need. Birdbaths are also fun and easy to make. They don't have to be fancy. A wooden bowl either on the ground or attached to a pole with strong glue will work. You can a buy birdbaths, too. If you choose to have a birdbath in your garden, the most important thing to remember is to keep the water clean. Pictured below are a few different types.

PEDESTAL BIRDBATH

GROUND BIRDBATH

HANGING BIRDBATH

BIRD FEEDERS

Different types of bird feeders attract different birds.

Here are seven feeders and the birds they attract. Try hanging one or more in your garden or on your balcony or patio. Try different types of treats, too. Birds eat a variety of seeds and nuts and even suet.

HOUSE FINCH

WINDOW FEEDER

Fill with: Seed mix or black oil sunflower seeds

Attracts: Chickadees, house finches, cardinals, sparrows, goldfinches, other small birds

NECTAR FEEDER

Fill with: Nectar (make it at home with one part white sugar dissolved in four parts water; no food coloring)

Attracts: Hummingbirds

RUFOUS HUMMINGBIRD

RED-BELLIED WOODPECKER

SUET HOLDER

Fill with: Suet

Attracts: Woodpeckers, flickers, nuthatches, brown creepers, wrens, chickadees, bluebirds, cardinals, mockingbirds

Tip: To keep big birds away, use a cage-within–cage suet holder.

EURASIAN BLUE TIT

TRAY OR BOWL FEEDER

Fill with: Nuts and seeds

Attracts: Ground-feeding birds, including doves, juncos, sparrows, blackbirds, jays

HOPPER FEEDER

Fill with: Nuts and seeds
Attracts: flickers, grosbeaks, jays

NORTHERN CARDINAL

TUBE FEEDER

Fill with: Nuts and seeds, or mealworms
Attracts: Cardinals, sparrows, grosbeaks, chickadees, titmice, finches

HOUSE FINCH

EUROPEAN STARLING

GLOBE FEEDER

Fill with: Seeds and nuts
Attracts: Goldfinches, chickadees, siskins

Bird Party

Invitations sent,
the feeders filled,
millet stocked,
and bread's been spilled.

The guests arrive
without delay,
robins, wrens,
one noisy jay.

In the grass
and on the deck,
doves mill around,
while finches peck.

The crows' attempted
party crashing—
met with vigilant
thrashers thrashing.

Chit chit, chit chit
chick-a-dee-dee-dee
It's a backyard bird
cacophony!

—Heidi E. Y. Stemple

181

AUTHORS' NOTES

JANE:

Professor David Stemple taught our family to bird. Born in West Virginia, he grew up in the mountains. When he graduated from college, he moved to New York City, where we met. He was the inspiration for my most famous bird book, *Owl Moon*. His birding pals called him a "lucky birder" because of the number of rarities he always managed to find.

"Not lucky at all," he would say. "I find those birds because I show up."

Showing up. It's a lesson our whole family has taken to heart. We show up. For the birds. For the work. For each other. It's what we do.

And that's also how this book was born.

JASON:

Growing up in a birding family meant seeing the world in a different way. It quickly became clear to me that the actions we take as humans have drastic effects on the ecosystems around us and of course the birdlife as well. It also meant that many of our travels and activities revolved around birds. When I was 10, we drove across the country from coast to coast, zig-zagging our way from one new bird to the next. And when I was 11 my father took me on a birding trip to the Everglades, where I fell in love, not only with the place, but with the idea of the wild and immersing one's self in it. The birds were always the driving factor and determined the destinations, but the rewards of a youth spent chasing birds went much deeper.

ADAM:

My oldest memories are of watching birds with my father. Whether it was finches at the feeders, warblers in the woods, or hawks soaring over the sunroof of the car, birds were as much a part of my childhood as going to school or playing with my friends.

My father passed away more than a decade ago. But every bird I see recalls the expertise and awe that he brought to birding, his love of the natural world, and the joy he felt in sharing that love with his family.

HEIDI:

I quit birding after my dad died. I just couldn't. But, one day, I was hiking when a bird flushed by the shoreline. "Ruddy turnstone!" I blurted out without thinking. And, indeed, it was. All those birds were still in my head. I started to look for them again.

Owling, though, I hadn't given up. My dad taught us all, but he and I kept owling for the Audubon Christmas Bird Count together until it was just me. On our best night, my dad and I had called down an impressive 34 owls. After he was gone, I formed a group called the OMG (Owl Moon Gang) and we called down 67. I know Daddy would be so proud. Of the owls, but also of this book. It is our love letter to him.

This is a huge book with lots of pieces. These are some of the people who helped make it possible.

Thank you to Don Kroodsma and Suzanne Shoemaker for their interviews and their amazing work with and knowledge of birds, as well as their good humor.

Thank you to Aiden Griffiths, young bird-watcher extraordinaire, for helping us old folks understand the new birding technology.

Thank you to Geoff LeBaron for vetting Heidi's original research for the Audubon Christmas Bird Count. Thank you to Greg Budney and all the Cornell Lab of Ornithology folks who have always helped the Stemple family when needed.

As always, thanks to Jane Wald of the Emily Dickinson Museum.

A huge hug to the Hampshire Bird Club (especially Janice, Jan, and Dan) as well as Monte and Bill, who are all co-conspirators in Heidi's quest for owls. Thanks for the compiling and the radio love.

Thanks to Priyanka Lamichhane and the National Geographic Kids crew who (literally) made this book happen. Thanks to Elizabeth Harding who held our (metaphoric) hands along the way.

And a special nod to the members of the OMG: Susannah Richards, Lynn Pelland, Brian Cassie, Sloan Tomlinson, Stephen Swinburne, and Dennis Wehrly.

FIND OUT MORE

Note: Please make sure to ask a caregiver or trusted adult for permission or to help you go online.

WEBSITES OR BOOKS ON TOPICS IN THIS BOOK

Watch Mongolian Eagle Hunters in action:
GoPro: Eagle Hunters in a New World
youtube.com/watch?v=WvhM8Lc9m-o
On the Western plains of Mongolia a nomadic group of Kazakhs continue the ancient practice of hunting with golden eagles.

Additional information about how to find and join bird counts:

Join the Christmas Bird Count:
audubon.org/conservation/join/Christmas-bird-count
Participate in the 117th Audubon Christmas Bird Count. With your help, the data will fuel important science and conservation work.

To find an Audubon chapter closest to where you live, visit:
audubon.org/audubon-near-you

If you would like to join a different bird count, there is a Great Backyard Bird Count every February:
audubon.org/conservation/about-great-backyard-bird-count
or gbbc.birdcount.org/about

Interested in counting the birds where you live? Join the Big Year bird count:
listing.aba.org/big-year-rules

Big Year sound like too much of a commitment? Try a Big Day bird count. Gather a team and see how many species you can see in a 24-hour period. You can find more information here:
listing.aba.org/big-day-count-rules

The Backyard Birdsong Guides by Donald Kroodsma, illustrated by Larry McQueen:

The Backyard Birdsong Guide Eastern and Central North America: A Guide to Listening. By Donald Kroodsma. Cornell Lab Publishing Group, 2016.

The Backyard Birdsong Guide Western North America: A Guide to Listening. By Donald Kroodsma. Cornell Lab Publishing Group, 2016.

Owl Moon Raptor Center:
OwlMoon.org

TO FIND WAYS OF JOINING OTHER CITIZEN SCIENTISTS, TRY THESE ONLINE SITES:

Citizen Science League:
cshl.libguides.com/c.php?g=474068&p=3243850

Project Feeder Watch:
feederwatch.org

Nest Watch:
americanornithology.org/content/citizen-science-projects

Urban Bird Research:
celebrateurbanbirds.org

SELECTED BIBLIOGRAPHY

Animals Charles Darwin Saw: An Around-the-World Adventure. By Sandra Markle. Chronicle Books, 2009.

Birds Do the Strangest Things. By Leonora and Arthur Hornblow. New York Step-Up Books, Random House, 1965.

Bird Watching Answer Book. By Laura Erickson. Cornell Lab of Ornithology, Storey Publishing, 2009.

Charles and Emma: The Darwins' Leap of Faith. By Deborah Heiligman. Henry Holt & Company, 2009.

Finding Wonders: Three Girls Who Changed Science. By Jeannine Atkins. Atheneum Books for Young Readers, 2016.

Firebird. By Jane Yolen. HarperCollins, 2002.

H is for Hawk. By Helen Macdonald. Grove Press, 2014.

National Geographic Field Guide to the Birds of North America, 7th ed. By Jon L. Dunn and Jonathan Alderfer. National Geographic, 2017.

The Peterson Field Guide Series: A Field Guide to Eastern Birds' Nests. By Hal H. Harrison. Houghton Mifflin Harcourt, 1998.

The Peterson Field Guide Series: A Field Guide to Western Birds' Nests. By Hal H. Harrison. Houghton Mifflin Harcourt, 2001.

Stone Girl Bone Girl: The Story of Mary Anning of Lyme Regis. By Laurence Anholt. Frances Lincoln Children's Books, 2006.

What's That Bird? By Joseph Choiniere and Claire Mowbray Golding. Storey Publishing, 2005.

OTHER BOOKS AND WEBSITES ABOUT BIRDS

Birds, Nests, and Eggs. By Mel Boring. Cooper Square Publishing, 1998.

Hello, World! Birds. By Jill McDonald. Doubleday Books for Young Readers, 2017.

National Geographic Kids Bird Guide of North America, 2nd Edition. By Jonathan Alderfer. National Geographic, 2018.

National Geographic Kids Ultimate Explorer Field Guide: Birds. By Julie Beer. National Geographic, 2016.

National Geographic Little Kids First Big Book of Birds. By Catherine D. Hughes. National Geographic, 2016.

For information on National Geographic's Year of the Bird initiative, visit the National Geographic Kids website:
natgeokids.com/animals/hubs/birds/year-of-the-bird

For Dr. Taylor's crow video, in which 007 solves an eight-step puzzle, visit:
wimp.com/a-crow-solves-an-eight-step-puzzle

CREDITS

For the Love of Birds: Jane Yolen

What Is a Bird?
What Defines a Bird: Jane Yolen
Bird Anatomy: Heidi E. Y. Stemple
Eggs: Jane Yolen
Nests: Heidi E. Y. Stemple
Nest Gallery: Heidi E. Y. Stemple
Beaks: Jane Yolen
Beaks Gallery: Heidi E. Y. Stemple
Bones: Jane Yolen
Wings: Jane Yolen
Wings Gallery: Heidi E. Y. Stemple
Wingspans: Heidi E. Y. Stemple
Feathers: Jane Yolen
Parts of a Feather: Heidi E. Y. Stemple
Feather Fun Facts: Heidi E. Y. Stemple
Feathers Gallery: Heidi E. Y. Stemple

Ancient History of Birds
Dino Birds: Jane Yolen

Birds in History
Man's Best Friend: Adam Stemple
Falconry in the Middle Ages and
   Beyond: Jane Yolen
Ravens of the Tower: Adam Stemple
Audubon Christmas Bird Count: Heidi
   E. Y. Stemple

The Peabody Ducks: Jason Stemple

State Birds
Jane Yolen and Heidi E. Y. Stemple

Listening to Birds
Birdsong: Jane Yolen
Listening to Wrens: Heidi E. Y. Stemple
Sonograms: Adam Stemple
How Do Birds Hear?: Jason Stemple

Looking at Birds
Birds Are All Around: Jane Yolen
Bird Tech: Heidi E. Y. Stemple
How to Photograph Birds: Jason
   Stemple
How Do Birds See?: Jason Stemple

Birds on the Move
Migration: Jane Yolen
How Do Birds Know Where They're
   Going?: Jane Yolen
Long Journeys: Heidi E. Y. Stemple
The Missing Swallows of San Juan
   Capistrano: Heidi E. Y. Stemple
Birds in Flocks: Jane Yolen

Saving Our Birds
Passenger Pigeon: Jane Yolen
The Dodo: Adam Stemple
Heath Hen: Jason Stemple

Ivory-Billed Woodpecker: Heidi E. Y.
   Stemple
Introduced Species: Heidi E. Y. Stemple
Endangered: Heidi E. Y. Stemple
Great Green Macaw: Heidi E. Y. Stemple
California Condor: Heidi E. Y. Stemple
Bali Starling: Heidi E. Y. Stemple
Whooping Crane: Heidi E. Y. Stemple
Cerulean Warbler: Heidi E. Y. Stemple
Success Stories: Heidi E. Y. Stemple
Bald Eagle: Adam Stemple
Raptor Rehabilitation: Heidi E. Y.
   Stemple
Osprey: Heidi E. Y. Stemple
What You Can Do to Help Raptors:
   Heidi E. Y. Stemple
Wild Turkey: Heidi E. Y. Stemple
Pied Avocet: Heidi E. Y. Stemple

Bird Records
Jane Yolen and Heidi E. Y. Stemple

Birds in the Arts
Birds in Print: Audubon's *Birds of
   America*: Jason Stemple
Birds in Music: Jane Yolen, Adam
   Stemple, Heidi E. Y. Stemple
Firebird, a Ballet: Heidi E. Y. Stemple
Birds in Poetry: Heidi E. Y. Stemple
Birds on the Screen: Jane Yolen
Birds in Story: Retold by Jane Yolen

Citizen Science

Famous Citizen Scientists: Jane Yolen

Becoming a Citizen Scientist: Jane Yolen

Birds in Your Backyard

All pieces in this section by Jane Yolen and Heidi E. Y. Stemple

POETRY AND SONGS

Poems

Two Eggs, One Goose © 2018 Jane Yolen

The Fallen Nest © 2018 Jane Yolen

A Beak © 2018 Heidi E. Y. Stemple

Hollow Bone © 2018 Jane Yolen

To Fly or Not To Fly © 2018 Heidi E. Y. Stemple

Shake a Tail Feather © 2018 Heidi E. Y. Stemple

Once A Week From the Paleontologist © 2018 Jane Yolen

Counting Birds © 2018 Heidi E. Y. Stemple

That Bird Song © 2018 Jane Yolen

On the Hunt © Jane Yolen & Heidi E. Y. Stemple

Bird Flock couplets all © 2018 Jane Yolen

Vee © 2018 Heidi E. Y. Stemple

Egg Shells: A Lullaby © 2018 Heidi E. Y. Stemple

Bird Olympics © 2018 Jane Yolen

The Bird in Audubon's Picture © 2018 Jane Yolen

Anna Pavlova's Swan © 2018 Jane Yolen

Do No Harm © 2018 Heidi E. Y. Stemple

Bird Party © 2018 Heidi E. Y. Stemple

Poems by Emily Dickinson: Public Domain. "The Robin" and "In The Garden" from *Poems* (Series Two), by Emily Dickinson. Edited by Mabel Loomis Todd and T. W. Higginson, through Project Gutenberg: http://www .gutenberg.org/cache/epub/2679/ pg2679-images.html .

Songs

I Love My Rooster: Traditional. Original musical arrangement by Adam Stemple. First published in *Jane Yolen's Old MacDonald Songbook*, Boyds Mills Press © 1994. All rights reserved.

Hush Little Baby: Adapted by Jane Yolen and Heidi Stemple © 2018. Tune is traditional. Original musical arrangement by Adam Stemple. First published in *The Lullaby Songbook*, Harcourt Brace Jovanovich © 1986. All rights reserved.

LIST OF SCIENTIFIC NAMES OF BIRDS IN ORDER OF APPEARANCE

INDEX

Boldface indicates illustrations.

PHOTO CREDITS

For David W. Stemple—Dov, Daddy, Dad, Papa, Pa—who
always showed up for the birds and taught us all to do the same.
This book is for you.

For librarians and teachers: ngchildrensbooks.org

More for kids from National Geographic:
natgeokids.com

National Geographic Kids magazine inspires children
to explore their world with fun yet educational
articles on animals, science, nature, and more.
Using fresh storytelling and amazing photography,
Nat Geo Kids shows kids ages 6 to 14 the fascinating
truth about the world—and why they should care.
kids.nationalgeographic.com/subscribe

For information about special discounts for bulk
purchases, please contact National Geographic
Books Special Sales: specialsales@natgeo.com

For rights or permissions inquiries, please contact
National Geographic Books Subsidiary Rights:
bookrights@natgeo.com

Designed by Julide Dengel

National Geographic supports K–12 educators with
ELA Common Core Resources. Visit natgeoed.org/
commoncore for more information.

Hardcover ISBN: 978-1-4263-3181-7
Reinforced library binding ISBN: 978-1-4263-3182-4

Printed in China
18/PPS/1

If it walks like a duck, quacks like a duck, looks like a duck, it must be a duck.

He takes to it like a duck to water.